IFLA

International Federation of Library Associations and Institutions
Fédération Internationale des Associations de Bibliothécaires et des Bibliothèques
Internationaler Verband der bibliothekarischen Vereine und Institutionen
Международная Федерация Библиотечных Ассоциаций и Учреждений
Federación Internacional de Asociaciones de Bibliotecarios y Bibliotecas

IFLA Publications 55

Continuing Professional Education: an IFLA Guidebook

A Publication of the
Continuing Professional Education
Round Table (CPERT)
of the International Federation of
Library Associations and Institutions

Edited by Blanche Woolls

Reports from the Field edited by
Miriam H. Tees

K·G·Saur
München · London · New York · Paris 1991

IFLA Publications
edited by Carol Henry

Recommended catalog entry:
Continuing Professional Education:
an IFLA Guidebook /
Edited by Blanche Woolls (Reports from
the field edited by Miriam H. Tees)
under the auspices of the Continuing Professional
Education Round Table (CPERT) of the International
Federation of Library Associations and Institutions
München, London, New York, Paris; K. G. Saur, 1991
158 p., 21 cm –
(IFLA Publications; 55)
ISBN 3-598-21784-6

Die Deutsche Bibliothek – CIP-Einheitsaufnahme

Continuing professional education : an IFLA guidebook ; a publication of the Continuing Professional Education Round Table (CPERT) of the International Federation of Library Associations and Institutions (IFLA) / ed. by Blanche Woolls. Reports from the field ed. by Miriam H. Tees. – München ; London ; New York ; Paris : Saur, 1991
(IFLA publications ; 55)
ISBN 3-598-21784-6

Ne: Woolls, Blanche [Hrsg.]; International Federation of Library Associations and Institutions / Continuing Professional Education Round Table; International Federation of Library Associations and Institutions: IFLA publications

∞
Printed on acid-free paper

K. G. Saur Verlag GmbH & Co. KG, München 1991
Printed in the Federal Republic of Germany

Druck/Printed by Strauss Offsetdruck GmbH, Hirschberg
Binden/Bound by Buchbinderei Schaumann, Darmstadt

ISBN 3-598-21784-6
ISSN 0344-6891 (IFLA Publications)

TABLE OF CONTENTS

PART I

THEORY AND PRACTICE OF CONTINUING PROFESSIONAL EDUCATION

PART II

CONTINUING PROFESSIONAL ACTIVITIES AND PROGRAMMES WORLDWIDE

APPENDICES

PREFACE

Continuing Professional Education: An IFLA Guidebook, an introduction to the preparation of continuing education programmes, is but another in the many products, often conceived and always implemented by Dr. Elizabeth W. Stone. She deserves full credit and acknowledgement for the execution of this activity. Through her leadership and management, the outline for the volume was determined. Authors were invited to write articles and gently prodded to see that they submitted the manuscripts. Finally, her efforts provided the impetus to the editing and final production of the work itself.

Articles are written to provide the reader with step-by-step instructions for planning and executing continuing education programmes. The activities are presented at a basic level and the suggestions can be implemented by the novice who wishes to conduct continuing education.

Part I begins with a thoughtful presentation of the *raison d'etre* for the *Guidebook.* Paul Kaegbein's summary of the CPE Seminar in Martin, Czechoslovakia held October 10-14, 1988 sets the stage for the articles that follow. John van Niekerk provides a general outline for planning continuing education and Fransie Terblanche explains the role of training in employee development.

Specific suggestions for designing distance learning are given by Laurel A. Clyde, by M. Kent Mayfield for tutorials as a strategy for learning through networking, and by Katherine Harig for poster sessions. C.M. Vink explains the role of the employer as trainer.

Using instructional technology in continuing education is outlined by Blanche Woolls. Darlene E. Weingand discusses methods for marketing professional education, and Margaret Gallacher reports the application of marketing techniques to a specific activity, a seminar programme.

Henry Blid and Lars Arvidson share the history and application of study circles in Stockholm. U.S. government support for continuing professional education is outlined by Anne J. Mathews and Thomas C. Phelps. Bruce E. Massis provides an update of library schools offering training courses in services to the print handicapped. Michel J. Menou provides the status of a general programme for education in agricultural information.

Two authors, Martha Boaz and Joseph Becker, give the reader a vision of the future. Boaz discusses anticipation of important changes in the library and information science profession, and Becker shares the concept of the University of the World.

One of the objectives of the *Guidebook* was to locate persons around the world who are willing to serve on a continuing basis as reporters for CPE developments in their geographic area. Those who submitted reports for this volume have been included.

Continuing professional activities worldwide, submitted by several countries, have been reported in Part II. These were carefully compiled and edited by Miriam H. Tees, Professor Emeritus, McGill University, Montreal, Canada, and Immediate Past Chairman of the Coordinating Board of the Division VII of IFLA, Education and Research.

The appendices provide two excellent bibliographies. Thelma H. Tate has listed continuing professional resources in the United States, and the results of an ERIC search has been cited. The IFLA Continuing Professional Education Round Table Mailing List, ably maintained since the CPERT first started by Mary K. Feldman, Library Consultant, can be obtained from Dr. Elizabeth W. Stone, School of Library and Information Science, The Catholic University of America, Washington DC 20024, USA. Finally, the Guidelines for Quality in Continuing Education for Information, Library, and Media Personnel adopted by the Continuing Education Subcommittee of the Standing Committee on Library Education (SCOLE) of the American Library Association are reproduced.

Appreciation is expressed to the writers of the papers and the reporters from each country, to Miriam Tees for her editing and to Mary K. Feldman for her efforts toward publication. In addition, appreciation is expressed to Dean Deanna Marcum, School of Library and Information Science, The Catholic University of America for permitting those associated with this project to use the equipment and facilities.

One final note: the *Guidebook* is perhaps more a "sampler" for it does not present articles on every facet of continuing education. Further, the volume perhaps reflects the same shortcoming that Paul Kaegbein stated in his summary, "...thelevel of the texts presented varied enormously, from precise reports of results on special research topics with data and sources cited, to experiences based on practice from individual institutions." However, it is the *Guidebook* requested by members of CPERT, and it will be members of CPERT who create the next *Guidebook*. It is up to the reader to inform CPERT of the needs.

Blanche Woolls, Editor
July, 1990

PART I

THEORY AND PRACTICE OF CONTINUING PROFESSIONAL EDUCATION

INTRODUCTION

Elizabeth W. Stone
Professor Emeritus and Former Dean
School of Library and Information Science
The Catholic University of America;
Co-Chair, 1st World Conference on Continuing Education for the Library and Information Science Professions; and,
Chair, IFLA CPERT

It was in the Czechoslovakian Alps during the International Federation of Library Associations and Institutions (IFLA) Conference of 1978 that the suggestion was made for the publication of a continuing education guidebook (1) that would present practical, pragmatic approaches to continuing professional education. It was emphasized that what is good for one institution, or one organization, does not necessarily mean that the same procedures would be appropriate in a different environment. However, examples of the programmes and activities presented offer support for any person who has been given an assignment of directing a specific continuing education activity covered in this *Guidebook*.

During the 1st World Conference on Continuing Education for Library and Information Science Professions held in Illinois just prior to the IFLA 1985 Conference in Chicago, and, also during the first meeting of the Continuing Professional Education Round Table (CPERT) in Tokyo in 1986, a questionnaire filled out by the attendees showed a need for practical information on continuing professional education (CPE) programmes and activities; the development of some way to share CPE ideas; and the information necessary to build a network of persons actively involved in CPE.

The need for such a publication was expressed even more specifically by participants in the IFLA CPERT programme meetings in Brighton in response to a questionnaire distributed in 1987. The following were the chief items the Brighton attendees wanted included:

- A manual with ideas on how to produce good CPE programmes;
- A clearing house for CPE information and materials;
- Resources that can be used by trainers of trainers;
- Descriptions of CPE programmes worldwide; and,
- Listing of those involved in CPE as an aid in networking.

In June, 1988, in response to a request from IFLA Headquarters asking for suggestions for major projects (for which some funds were available), an *ad hoc* group of IFLA CPERT members met with Brooke Sheldon (co-Chair of the 1st World Conference on Continuing Professional Education) during the ALA New Orleans Annual Conference. This group recommended that the next special project of the CPERT be a *Continuing Professional Yearbook* and that this resolution be presented during IFLA Sydney Conference at the CPERT Executive/Planning meeting. Accordingly, on August 30, 1988, this proposal was discussed and it was the consensus that such a proposal be presented to the IFLA Professional Board for approval. It was, however, recommended that the title be changed from *Yearbook* to *Guidebook*.

At this Sydney programme meeting of the CPERT, the nature of the proposal was outlined and approved, and the call went forth for reporters to volunteer to cover CPE activities on a regular basis in their geographical area. At this meeting,

21 attendees signed up to help with various phases of the *Guidebook*, and most of these are contributors to this volume!

The next step was submitting a formal proposal to the IFLA Professional Board. It was stated in this proposal that the purpose of the *Guidebook* was to bring together current, authoritative, enduring information on continuing professional education topics as one way of fulfilling the "Terms of Reference" for the CPERT.

These terms of reference are:

- Provide and advance continuing professional education or personnel in the library and information fields; and,
- Provide a forum for the exchange of ideas and information on all aspects of continuing professional education.

The proposal emphasized that the publication would be a systematic and efficient way of implementing many of the items in the Medium-Term Programme (MTP) 1986-1991. (2) As stated in the MTP, these eight action plans of the CPERT are:

1. Promote continuing education for library and information personnel as a significant part of local, national, and international library and information science efforts.
2. Identify elements that yield quality Continuing Education and ways to achieve these elements.
3. Study alternate delivery systems for Continuing Education and their relative effectiveness.
4. Bring together on a regular basis leading scholars, librarians, and library educators from around the world as a means of facilitating collaboration and the development of networks for the exchange of ideas through correspondence, newsletters, scholarly papers, and meetings.
5. Serve as a clearing house for Continuing Education achievements, needs, problems, financial support and other possible solutions.
6. Improve the ability of librarians worldwide to contribute to the lifelong learning of individuals both within and outside the profession.
7. Encourage international exchanges for library personnel; and identify institutions, professional associations and other organizations willing to facilitate exchanges and/or make available resources to support this activity.
8. Encourage IFLA to influence governments and organizations at the regional, national, state and local levels and to increase their understanding of the need to support and/or make funds available for the education of library personnel at the basic and post-qualification levels emphasizing the real value of this investment to national development.

In this first edition of the *Guidebook*, the papers were all sent in by people responding to a "Call for Papers". Readers can choose what would best fit their particular needs. Readers should feel free to contact any of the persons who have contributed to the *Guidebook* for more details relevant to the programmes and projects described.

References

1. Stone, Elizabeth. *IFLA and the Continuing Education of Librarians: A World Perspective.*Paper presented at the IFLA Conference in Czechoslovakia. 1978.
2. 3.7.6 Round Table on Continuing Professional Education: In: *Medium-Term Programme 1986-1991*,2d ed. The Hague, Netherlands: IFLA. 1988, pp. 59-60.

NEW DIRECTIONS IN CONTINUING PROFESSIONAL EDUCATION FOR LIBRARIANS CONFERENCE REPORT: CPE SEMINAR IN MARTIN, CZECHOSLOVAKIA

Paul Kaegbein
Lehrstuhl für Bibliothekswissenschaft
der Universität zu Köln, Universitätstrasse 33, Köln, Germany

ABSTRACT

A summary of the CPE Seminar held in Martin, Czechoslovakia, October 10-14, 1988 on the occasion of the 125th anniversary of the Matica Slovenska and the tenth anniversary of the continuing education center for Slovak librarians.

At the 1977 Jubilee Conference of IFLA in Brussels, there was interest expressed in forming a special group of experts dedicated to continuing professional education. One year later, Elizabeth W. Stone gave a fundamental talk on this topic at a session of the Section of Library Schools during the IFLA conference in Strbske Pleso (1). Both can be seen as the beginning of a development which, after lengthy preparations, first led to the 1985 *World Conference on Continuing Professional Education for the Library and Information Science Professions* in Chicago (2-4). In 1986, the IFLA Continuing Education Round Table (IFLA CPERT) was founded within the renamed IFLA Section for Education and Training.

This Round Table has since developed into a noteworthy activity at the international level. Its bulletin, *Update* was complemented with a *Newsletter* in 1987, 450 copies of which are already being sent to 60 countries. [The *Newsletter* continues at the present time.] The *International Directory of Continuing Education Specialists in the Library and Information Science Professions*, which has been prepared by the Round Table, will be a useful aid for future contacts between professors active in the area of continuing education for librarians.

The occasion of the tenth anniversary of the continuing education centre for Slovak librarians at Matica Slovenska in Martin in the same year in which Matica Slovenska, this "Mother of Slovakia," was celebrating 125 years since its founding, (5) was used by the IFLA Round Table to organize an international seminar on questions concerning continuing education for librarians, held on October 10-14, 1988. More than 40 participants gathered, including a total of 16 from Canada, Germany, Great Britain, Hungary, Poland and the USA. The conference was opened by the director of the Slovak National Library in Matica Slovenska, Miroslav Bielik. Inaugural speeches stressing the meaning of continuing education for librarians were made by IFLA President Hans-Peter Geh; Helena Palkova, the representative of the Ministry of Culture of the Slovak Socialist Republic in Bratislava; Elizabeth W. Stone, the chair for the IFLA Round Table; and Horst Hogh, the director of the continuing education courses at Matica Slovenska, who was also responsible for the organization of the seminar.

The seminar lectures were assigned to seven topic groups, according to their contents. The first one dealt with the basic relationship between professional and continuing education in the area of librarianship. Jifi Kabrt (Institute for Library Science and Scientific Information, Charles University, Prague) presented in his report, "On mutual relations between the system of school education and the system of further education of librarians," the government measures which have been introduced in Czechoslovakia since 1973, which aim at a general equality of professional and continuing education in all fields. The standard framework is also to be filled out in the realm of librarianship. Just how such a task could be carried out was investigated by Milada Ulbrichova (Centre for Scientific, Technical and Economic Information, Prague) in her lecture, "On relations between the system of school education and the system of further education in the sphere of scientific information and librarianship in the Czechoslovak Socialist Republic." During the discussion, Eugen Messa (Institute for Library Science and Scientific Information, Comenius University, Bratislava) added her report on the relationship of this institute to the Matica Slovenska in the field of continuing education of librarians, concerning which relevant publications, such as the journal *Citatel,* as well as the engagement of professors in the continuing education seminars can be cited.

The technological developments visible in all areas of information, especially at present, and which are partly responsible for considerable changes in job outlines, provided Beth Miller (School of Library and Information Science, Elborn College, University of Western Ontario, London, Ont.) with an opportunity to formulate "Educational challenges for the survival of librarianship." She argued above all for a basic, professionally useful training of future librarians as well as for a balanced relationship between theory and practice. Building on such a training programme, it should then become the task of continuing education to distribute specialized knowledge and directly applicable skills.

In the second topic group, which was dedicated to the future tasks of continuing education in relation to the development of librarianship, Elizabeth W. Stone (Dean Emeritus, School of Library and Information Science, The Catholic University of America, Washington, DC) offered clearly distinct concepts in content and form for continuing professional education for librarians in a broadly based lecture, "The quest for quality in continuing education," in which continuing education was clearly distinguished from preparatory professional education. She stressed especially the various learning strategies for children and adults as well as the individually differentiated forms of independent study for these groups; an effective continuing education programme for librarians must take them into account in terms of layout and content.

Horst Hogh (Martin) also proceeded from theoretical considerations. He investigated "Starting points for appointing goals and content of continuing professional education" in the light of conditions in Slovakia; thus, for example, continuing education programmes for staff members must be worked out and constantly run in every library, under legal regulations. For centrally planned activities, including those of the Matica Slovenska, he listed those topics which must be urgently considered in the near future, in conjunction with the changing tasks of libraries; the list includes, among other things, foreign languages, new media and the preservation of library materials.

Josephine Riss Fang (Graduate School of Library and Information Science, Simmons College, Boston, MA) raised the importance of questions on preservation matters as a topic in the third group of contributions, which dealt with specific areas of content in continuing education for librarians. In her lecture, "Continuing professional education in preservation management for libraries and archives," she analyzed the sections considered here according to groups of aims and administrative measures, and she argued for equal treatment of preservation in comparison with traditional methods of managing information materials, which applied both to continuing as well as professional education. This approach had become part of the regular training offered at Simmons College, after the first continuing education seminars on preservation had met with extraordinarily great interest.

Zita Musutova (Central Library of the Information Centre of the Slovak Academy of Sciences, Bratislava) shed some light on the structure peculiar to all academies of sciences in socialist countries, which also encompasses the librarian facilities connected to special library systems, in her discussion on "Starting points and criteria for selection of specializations and for stating the content of education in the library information system of the Slovak Academy of Sciences." She explained the principles which the library follows to fulfil its continuing education task. This presentation was complemented from the angle of public libraries by Elena Vel'asova (Municipal Library of Bratislava) in her contribution, "Starting points and criteria for selection of specializations and for stating the content of education." This referred above all to the various qualifications of the staff members in a library, which must be taken into account in the content, structure and level of continuing education seminars.

The fact that more than 60% of the loans from Slovak public libraries in 1987 involved belles-lettres, offered Stefan Kolivosko (Regional Library of Kosice) the opportunity to speak about the continuing education demands in this area ("On some questions concerning librarians' work with belles-lettres"). Good literature can help combat in part the generally observable retreat of humanistically oriented school education and thus guide the reader to special areas of art and culture. The speaker demonstrated the problems facing the librarian of a public library in his role as reference librarian, but he did not proffer any solutions for continuing education measures. These are also, according to Tamas Szepesvary (Institute for Librarianship, Pedagogical Institute, Eorvos Lorand University, Budapest) indispensable in the area of bibliographic research methods due to the strong rise in the number of bibliographic databases with varying retrieval methods ("Continuing professional education (CPE) in bibliographic services").

The next group was devoted to the didactic questions raised by continuing education. Mihaly Palvolgyi (Institute for Librarianship, Pedagogical Polytechnic Daniel Bersenyi, Szombathely) gave attention to the "Methodological foundation of the continuous education by the primary professional education." Using his own institution as an example, he presented a model for instruction which could be used not only for the continuing education programmes of librarian-training institutions, but also for basic training. He established the theoretical and practical framework of the requirements for such a programme.

Marie Konigova (Institute for Library Science and Scientific Information, Charles University, Prague) gave a report based on her own years of teaching experience. Her lecture, "Some didactic aspects in the field of librarianship and scientific information," was based on motivating the participants at training schools to pursue independent study, above all in the investigation of textbooks with an eye to the didactic viewpoints which should result in the selection of texts suitable for continuing education programmes. Methods for such evaluations have been published by Bjornsson (6) and Nestler (7).

In her contribution, "Intensive developing course for school librarians," Gyula Toth (Institute for Librarianship, Pedagogical Polytechnic Daniel Berzenyi, Szombathely) reported on the continuing education courses for school librarians recently established at her institution. A 1987 ordinance from the Hungarian Ministry of Culture led to the integration of continuing education for school librarians into the general continuing education system for teachers, a forward-looking decision not only in terms of the qualifications gained by those who successfully completed such a course, but also in terms of the closely knit tasks related to literature and libraries which have to be carried out by teachers and librarians responsible for school libraries.

The discussion for these groups revolved mostly around the methodological problems of continuing education, behind which the organizational questions retreated. It appears, however, to be easier to find answers for the latter than to solve the former, that is, to achieve a consensus on the optimal procedures. Here lies an expansive and largely unordered field for methodological and didactic investigation, to which international research in continuing education in librarianship ought to pay attention.

The fifth topic group concerned questions of management for continuing education institutions. Ian M. Johnson (College of Librarianship Wales, Aberystwyth) explained the relationship between "Marketing and the managing of continuing professional education in librarianship and information science." This included basic questions concerning marketing theory and market research as well as the planning and promotion of professional continuing education together with the consequences for optimal management. The College of Librarianship Wales (now the School of Librarianship and Information Studies of the University of Wales) has comprehensive experience in the organization and running of continuing education seminars in the shape of the 16-year-old "International Graduate Summer Schools," and other British training programmes are also very active in this area (8). The aspects touched on here must be carefully considered for continuing education programmes if they are to be prepared successfully.

Among the most interesting developments in this regard is "The Newcastle experience of developing continuing education," which was described by Edna Blackie (Department of Librarianship and Information Studies, Newcastle upon Tyne Polytechnic). Characteristic of this mainly successful experiment was the attempt to integrate lesson units of basic and primary studies into a modular structure, within the framework of the flexibility in university studies required by law in Great Britain. Thus the individual need for continuing education could be met by participation in specific courses. This takes place formally via the use of the *Associate Student Scheme*, especially useful for those who are already professionally engaged. The scheme allows access to regular university classes for those interested.

Marta Fratricova (University Library of Bratislava) discussed "The role of managers of libraries in provision of continuing professional education," in which she went into particular detail on the already mentioned legal standards as formal presuppositions and on the specialized priorities of colleagues at the library. These aspects had to be considered with a view to maintaining good employee relations during the implementation of the continuing education measures; nor should the evaluation of such activities be ignored.

The possibilities for achieving greater efficiency in continuing education by means of didactic aids for content unity, the topic of the sixth group, were interpreted by M. Kent Mayfield (Medical Library Association, Chicago, Illinois) in his instructive presentation "Breaking the barriers to learning." The adult learner has been the focal point over the last two decades of numerous andragogic investigations, whose results were summarized by the speaker, who derived four strategies for the adequate transferral of knowledge: *Self-directed learning, synergogy, competency-referenced learning systems, transfer-of-training*, which he explained with the use of examples. A further text by M. Kent Mayfield, "Organizational change in continuing education," was not presented, but rather distributed for general information. It refers to the prehistory of the institutionalized continuing education programme of the Medical Library Association and the changes which it has undergone in the course of three decades. "Some problems of the continuing education of librarians and information specialists in Poland" was presented by Maria Brykczynska (Public Library of Warsaw) in an overview which also considered historical developments since 1918.

The last topic group concerned the future organizational structures of continuing education. Referring to the example of Nicaragua, Carmela Ruby (Library Consultant, Sacramento, CA) discussed "Continuing library education in developing countries: one approach to planning"; she summarized the results of a field study and pointed out problems of desirable and necessary partnerships between industrialized and developing nations during the planning and implementation of librarian continuing education programmes. Her models contained a series of basic points to consider in such an undertaking, which could also apply to other countries.

Anastazia Carska (Institute for Continuing Education for Staff Members in Cultural Institutions, Bratislava) touched in her report on general questions of the modernization of continuing education with particular reference to continuing education for librarians. She emphasized the as yet unsatisfactory external situation in the Slovak libraries - scarcity of teachers, of rooms and of didactic-technical equipment - a situation which can have thoroughly negative effects on the degree of success of the programme. The situation in Poland was highlighted by Jozef Lewicki (Centre for Continuing Education for Librarians, Warsaw) with "A few remarks on the system of education, additional training and professional improvement training of library and scientific information centre staff in Poland." The centre was founded in 1976 and since then has opened 15 branches across the country, where special continuing education courses take place alongside two-year courses leading to supplemental formal qualifications for untrained staff members in libraries. Special attention was given here to school libraries.

Jana Sodomkova (State Library of the Czech Socialist Republic, Prague) complemented this topic with her presentation "Library education in public libraries network in CSR: the tasks of the CSR State Library at securing the system of continuing professional education." The problems of carrying out continuing education programmes for library automation were raised by Gyorgy Sebestyen (Institute for Library and Information Science, Faculty of Arts and Philosophy, Eotvos Lorand University, Budapest) with "Some organizational experiences of continuing professional education in library automation," in which he was primarily concerned with the congruency of content and external form.

The final presentation was an interpretation of the effective guidelines for the continuing education of librarians in Slovakia, given by Horst Hogh (Martin) in his report "Methods in the organization of continuing professional education in the Slovak Socialist Republic for the 90 years." At the same time, this presentation offered an explanation of the continuing education activities of the Matica Slovenska, the central institution of librarianship in the country. The text "History of librarians' education" distributed at the seminar, which was written by Horst Hogh and his colleagues at the Matica Slovenska, can be considered a complement to this. It also contains sections on the continuing education for librarians in Slovakia.

Most of the reports mentioned here were already available in English at the beginning of the conference, so that the work of the simultaneous interpreters, (the official languages were English, Slovak and Czech), was made a good deal easier. In those instances where only a summary was presented due to time constraints, participants could still follow reports given in languages other than their own by referring to the translated texts. Nevertheless, it was noted that the number of presentations was too large and their assignment to a specific topic group seemed at times to depend on pure chance. The multitude of topics raised from the various countries also hindered the discussion. It would have been better if all the presentations from Czechoslovakia had offered, as most others had, references to the literature, to enable the reader to pursue individual aspects independently. In addition, the level of the texts presented varied enormously, from precise reports of results on special research topics with data and sources cited, to experiences based on practice from individual institutions. It was also apparent that the terminology was differently interpreted at times, which on top of the language barrier made it even more difficult to understand all meanings of the papers. This aspect should be considered when evaluating the texts presented.

Overall, continuing education for librarians, if considered from a methodological stance as a developing area of science, in which new findings constantly cause the revision of paradigms fixed earlier, is still in a relatively early phase of its development (9). Internationally recognized paradigms are still quite rare in this field. Nevertheless, as the previously mentioned *World Conference on Continuing Professional*

Education for the Library and Information Science Professions shows, the community of scientists interested in the problems of continuing education will not be able to avoid long-term requirements for greater precision, such as commonly understood terminology, if they are to reach a consensus (10). In this regard, the worth of such a seminar as the one in Martin, which led to the vitally necessary contact and exchange of thoughts between theoreticians and practitioners of continuing education (at least in certain areas) cannot be rated highly enough. Apart from the weak points already mentioned, there was an abundance of important information on practically tested or planned continuing education activities in various countries, which could lead to further discussion and reflection on one's own viewpoints in the light of methodologically oriented principles. This is especially valid for the necessary confrontation with the theoretically-based presentations to ensure the optimal formulation of continuing education programmes.

These viewpoints are also to be found in the recommendations of the participants at the end of the conference which will be published in full in the *IFLA Journal*. Not only are further seminars planned for the coming years on the problems touched on here under the aegis of the IFLA Round Table for Continuing Professional Education; there is also a need for this at the international level, as discussions showed, for the clarification of the fundamentals and for closer cooperation in practice. In addition, the development of continuing education guidelines for libraries is considered important at the national level. All relevant continuing education activities should be regularly brought to the attention of the IFLA Round Table, which in this way can function worldwide as the inquiry point and information centre for this area of research. This would also include supporting the general continuing education, and spurring further research.

In the closing meeting, Horst Hogh raised both the points of interest discussed in the presentations and the numerous problems in continuing education for librarians which still have to be solved. Words of thanks were extended by Elizabeth W. Stone and Hans-Peter Geh to the Matica Slovenska, the Slovak Ministry of Culture and the city of Martin, which made the seminar possible, and whose generously planned framework programme gave the foreign participants insights into the cultural specialties of Slovakia and the countryside surrounding Martin. Publication of the seminar presentations in English is planned, the publication can be obtained by contacting the IFLA Round Table on Continuing Professional Education.

References

1. Stone, Elizabeth E. *IFLA and The Continuing Education of Librarians: A World Perspective*. Unpublished paper available in English and French from the IFLA Clearinghouses (see IFLA Directory 1988/89). The Hague: IFLA. 1988, pp. 193-194.
2. Horne, Esther E., ed. *World Conference on Continuing Education for the Library and Information Science Professions. Continuing Education: Issues and Challenges.* Papers from the Conference held at Moraine Valley Community College, Palos Hills, IL., August 13-16, 1985. New York, Munich, London, Paris: Saur. 1985.
3. Bock, G. Conference report. *Zeitschrift für Bibliothekswesen und Bibliographie*. 33. 1986, pp. 184-188.
4. Brykczynska, M. Conference Report. *Przeglad biblioteczny* 54. 1986, pp. 202-205.
5. Kukuckova, E. [The Matica Slovenska in our National and Cultural History: An Annotated Bibliography. 76 titles from the years 1873-1988].
6. Bjomsson, C.H. *Lasbarhet*. Stockholm: Liber, 1968.

7. Nestler, K. [Investigations into the Form and Effect of Teaching Texts with Regard to the Lexical-SyntacticalAspects.] *Informationen zu Schulbuchfragen* 24. 1987, pp. 9-52.
8. Johnson, Ian M. The development of continuing education in British schools of librarianship and information studies. In: *Theorie et pratique dans l'enseignement des sciences de l'information.* Montrael: Ecole de Bibliotheconomie et des Sciences de l'Information, Universite de Montreal, 1988.
9. Kuhn, T.S. [The Structure of Scientific Revolutions.] Frankfurt am Main: Suhrkamp. 1967.
10. *Ibid.*

PRACTICAL HINTS WHEN PLANNING FOR CONTINUING EDUCATION

John van Niekerk
Deputy Director of Library Services
Medunsa University Library, Medunsa 0204, Southern Africa

ABSTRACT

The main steps in planning continuing education activities are discussed. Practical suggestions and examples are given which may assist in ensuring that these activities are presented in a well organized and purposeful manner.

The Process of Continuing Education

The following major aspects can be identified in the process of continuing education:
- Determination of learning needs;
- Description of learning objectives;
- Determination of learning content, materials and methods; and,
- Evaluation and recognition of learning.

This process is ongoing and the experiences and results of a continuing education activity will have an influence on future activities.

Learning Needs

Continuing education occurs not because it is generally agreed that it is "good" to acquire additional skills and knowledge, but because there is a need for this knowledge. In some cases, there may be a request from an individual or group for training in a specific area. Learning needs can also arise from inferior production and quality of work, e.g. a supervisor may see that certain staff are deficient in some respects and require additional training. New developments in a field may require that staff receive training.

Once a learning need has been identified, it is necessary to gather more information about it. This can be done by means of observation, discussions with staff at various levels as well as questionnaires and tests. In this way it is possible to determine what the learning need is and who requires the training. It is important that the persons for whom the training is intended as well as their supervisors should be involved in the determination of the learning needs so that a positive attitude towards the continuing education activity can be developed.

Learning Objectives

Once a learning need has been identified and described, a concise and clear statement of objectives for a continuing education activity which will result in the need being satisfied should be compiled. All aspects of the activity are then geared to the attainment of these objectives. At the end of the activity it must be possible to determine what participants have learned in terms of knowledge and skills and to what extent the objectives have been achieved. A continuing education activity may have a general objective as well as more specific objectives describing in more detail what is hoped to be achieved.

Example

General Objective. The librarian will be able to help users find information in a medical library.

Specific Objectives. The librarian will:

- be able to use an accepted reference technique to answer queries;
- know what the most common queries in a medical library are;
- know what the most important medical sources are and how to use them; and,
- be able to do literature searches in various periodical indexes.

Learning Content, Level of Presentation, and Programme

The learning content, materials, and methods of the continuing education activity as well as the resources and facilities used to present the programme all facilitate the attainment of the learning objectives. Care in planning at this stage increases the potential success of the participants.

The knowledge and abilities of participants should be taken into account in deciding the level of presentation. The level at which the learning is presented should not be too low or too complicated.

In planning the programme, topics should progress in a logical fashion to facilitate understanding and should be given concise but descriptive titles. Lectures or papers should not be too long to ensure that the audience remains interested. In most cases, 30 to 45 minutes should be sufficient.

The activity can be for part of a day, several days, or even a week or two depending on its nature. A workshop to discuss common problems may be for only one morning while an intensive course on a complex topic may require a week or longer. It is important, however, that participants be convinced that their time is likely to be well spent.

Speakers

Speakers should be knowledgeable in their fields and recognized as such by their profession. People may be hesitant to attend a continuing education activity if there is doubt about the credentials. Speakers should also be able to communicate their knowledge effectively.

Arrangements regarding speakers should be finalized well in advance as last minute arrangements place unnecessary pressure on both the speaker and organizers. Backup alternatives should be considered in case a speaker is unable to attend.

Instructional Methods, Aids, and Exhibits

If possible, a variety of instructional methods and aids should be used to promote the effective transfer of knowledge, e.g., a lecture supported by audiovisual material and followed by a practical example or discussion. Demonstrations and tours can also be arranged, preferably during the afternoon when delegates' concentration may be less intense.

Exhibits related to the theme of the continuing education programme can be mounted near the venue. It may be possible to negotiate sponsorship by private firms in return for the opportunity to exhibit products. Exhibits can provide valuable additional information to participants and introduce them to new development and products.

Evaluation

Provision should be made for the evaluation of a continuing education activity. Through thoughtful evaluation by participants and presenters it is possible to determine whether the learning objectives were achieved and to enable improvements in the future.

Recognition

While continuing education does not necessarily result in an academic qualification, some form of recognition can serve as encouragement of participants. One example of this is a certificate or statement indicating that a person attended a certain activity and what subjects were covered.

Administrative Matters

Various administrative aspects require attention when planning continuing education activities. If there is not efficient organization and administration, the transfer of knowledge to the participants is likely to be affected and the objectives of the activity will not be attained.

Organizer

A competent person should be appointed at an early stage to plan and implement a continuing education activity. Such a person should be able to:

- Plan logically and thoroughly;
- Supervise and communicate effectively with people;
- Control many different aspects of a situation and work efficiently under pressure; and,
- Make quick decisions if necessary and offer practical solutions to problems.

Depending on the scope of the activity, this organizer or planner may have several assistants.

Budget

It is important to estimate what the various costs of the continuing education activity will be so that a reasonable registration fee can be determined. Costs to bear in mind when planning are:

- Hire of the venue;
- Refreshments;
- Secretarial services, stationery, and printing;
- Advertising;
- Travel and accommodation, e.g., for speakers;
- Speakers' fees; and,
- Audiovisual equipment.

Income can be derived from:

- Sponsorship by private firms or other institutions such as a professional association; and,
- Registration fees. The costs involved in presenting the continuing education and the likely number of participants will be an important consideration when determining a suitable fee.

Date

A suitable date for the continuing education activity should be decided on well in advance. Holiday periods, unless they are for some reason convenient, should be avoided, as well as clashes with other activities in which participants may be involved. The coordination of several activities, functions, and meetings over a short period of time may sometimes be convenient.

Venue

A venue should be chosen which is convenient for the participants, not too far from transport and accommodation. The institution presenting the continuing education should have suitable premises. Alternatively, use can be made of a specialized conference centre or a hotel. While some hotels have good conference facilities, others adapt premises. When this happens, some problems can be encountered.

The venue should be reserved well ahead of time, as long as several months before the continuing education experience. It should be ensured that:

- The general appearance of the venue is neat and attractive.
- There is adequate, comfortable seating.
- Ventilation in the room or hall is good, preferably air conditioning.
- The acoustics are good and a public address system is available if required.
- Necessary audiovisual equipment is available or can be installed.
- Staff are available to assist where required, and their service is efficient. There should preferably be a single person who can be contacted should problems arise or help is needed.

Catering

A hotel will be able to supply the necessary tea, coffee, snacks, meals, etc. if the continuing education activity is held there. Some specialized conference centres may also have catering services available, but at other venues use will have to be made of outside catering services or the organizers will have to do their own catering. Menus and prices should be carefully checked and finalized well beforehand. One of the organizers could be delegated to handle catering matters and serve as liaison with the caterers. A few days before the continuing education activity, the hotel or caterers should again be contacted to confirm that the catering arrangements are in order. On the day of the training, it should be ascertained that refreshments and meals are served on time.

Advertising

The continuing education activity must be advertised so that potential participants are aware of it. Information should be conveyed indicating what the purposes of the activity are and how it can benefit the participants. The date of the activity, a preliminary programme and details of the speakers should also be given. A reply form should be included which people who wish to attend the activity may return as their reservation.

Advertising may take many forms. One of the more popular is the brochure. These may be folded so that they can be sent without the additional cost of an envelope. Care should be taken that full information is given and that information is not deleted when the reply form is torn off to return to make the reservation.

Professional associations may be willing to include advertising material in their routine mail to members. It may also be possible to place an advertisement in the newsletter or journal of such an association.

An address list of the institutions and people who may be interested in the continuing education activity should compiled. It may be possible to get addresses from branches of professional associations or other relevant bodies. A microcomputer can be very useful for storing the addresses and printing them on envelopes or letters when required.

Confirmation of Participation

After reply forms have been returned by prospective participants, standard letters should be sent to them confirming their participation and giving essential information such as the suggested travel route, parking, or limousine service from the airport to the venue. Receipts for money received should be attached to these letters.

Papers

For some continuing education programmes, papers and abstracts may be required beforehand for printing and inclusion in folders to be given to participants. Speakers should be requested to supply their papers a considerable time in advance, e.g. two months, as some will tend to be late.

Folders

For many continuing education activities, folders with stationery and pens may be desirable and are often available from private firms. A programme, list of participants, abstracts of papers, and any other necessary material can be included in the folders.

Name Tags

When a considerable number of participants are involved and the continuing education activity continues for a substantial period of time, participants should be identified so they may be addressed by other participants and the presentors. Name tags are useful if names and participants' institutions are given in clear, readable lettering.

Registrationof Participants

For a lecture or discussion, those attending may just gather in a room. For a more formal activity such as a conference, registration may be necessary. The followingshould be available:
- An alphabetical list of participants.
- A receipt system for handling money, and sufficient change.
- Sufficient registration officials and tables in the entrance to the venue.

When delegates arrive, their names can be ticked off on the registration list and they can be given folders, name tags, and any other pertinent information.

Assistance to Speakers

A person should be available to assist speakers with audiovisual equipment and any problems that may arise. A chairman should be appointed to introduce speakers and generally supervise proceedings in the venue. This person must ensure that speakers do not overrun their allocated time. Speakers and others who assisted with the continuing education activity should be thanked in an appropriate matter either personally or by letter.

Attention must be paid to many aspects when a continuing education activity is planned. If there is a deficiency in even a small detail, it can affect the success of the activity. It is therefore very important to plan carefully and well in advance to ensure that the activity proceeds smoothly and to the maximum benefit of the participants.

Additional Information: Council on the Continuing Education Unit. *Principles of Good Practice in Continuing Education*. Silver Spring, South Africa: The Council, 1984.

THE ROLE OF TRAINING IN EMPLOYEE DEVELOPMENT

Fransie Terblanche
Senior Lecturer
Department of Library and Information Science
University of South Africa, Pretoria, South Africa

ABSTRACT

The appropriate training of employees is regarded as one of the most important functions of effective human resources management. The purpose of training is to promote greater efficiency of employees in their current positions and to develop employees with a view to additional or other responsibilities. The creation of opportunities and facilities for professional growth is the shared responsibility of individuals, the profession, professional associations and library schools. The successful achievement of an organization's objectives should be the ultimate goal in any organization's strategy towards training and development. The organization's short-term and long-term objectives should be analyzed. In addition to this, the planning, development, and maintenance of human resources in any library and information organization should be efficient.

Training and development play a significant part in bringing together that which employees have to offer and that which the organization requires of them. The analysis of training needs as well as the advantages of training and the evaluation of training activities are highlighted in this article.

With due consideration to a developing South Africa and a changing library and information profession, it is clear that the training and development of library and information workers has become a specialized field. The training of employees is regarded as one of the main functions of efficient human resources management in both the public and private sectors.

Skilled and motivated library and information workers are a particular asset to a library and information service, but competence and motivation do not develop spontaneously. Effective training and development are needed to attain the desired level of proficiency.

A Definition of Training and Development Terminology

A distinction should be made between the following four concepts:

1. *Education* can be defined as the activity which is aimed at the development of knowledge, moral values, and understanding needed in all circumstances of life, rather than at knowledge and skills pertaining to a restricted field of activity. (1)
2. *Instruction* is aimed at expanding general knowledge in order to elucidate and understand certain phenomena. Instruction takes place at academic institutions before workers are employed, in order to prepare them for training. (2)
3. *Training* is defined as the systematic process by means of which the behaviour of employees is changed to the direction required to achieve the objectives of the organization. (3)
4. *Development* is described by one author as "...the systematic process of education, training, and growing by which a person learns and applies information, knowledge, skills, attitudes, and perceptions." (4) A second author defines development as "...skill development and the changing of attitudes among productive and operative workers." (5)

To summarize the points mentioned above, it can be said that training and instruction eventually culminate in development. Development is merely the individual's pursuit of personal aspirations after education has been completed. Individuals cannot develop, unless they have the intrinsic need to do so.

Continual education and further professional development are often described as *non-formal training* in the interrelationship with formal training (which includes education). This interrelationship is possible owing to the flexibility of non-formal training. Whereas formal training is structured with regard to venue, time, contents, presentation, and management, the flexibility of non-formal training in this respect has the advantage of being highly adaptable in providing for the rapidly changing needs and shortcomings that might be manifested in the professional preparation of the worker.

Non-formal training also functions as a substitute for formal training. Non-formal training provides opportunities for individuals to undergo training and take part in discussions on subjects or skills that were not covered by their initial basic training. By means of non-formal training, the participants' previous preparation for practicing their occupations can be continually refined.

Non-formal training also compensates for deficiencies and shortcomings in the participants' formal training. In this case the courses given in non-formal training, which is applicable to the practical labor situation, are of special importance.

Non-formal training also supplements formal training. The rapidly changing occupational labor situation requires workers to be up-to-date on the latest developments in their fields. The profession has to cope with the changing needs and requirements of its environment. To succeed in this, library and information workers need to update their own ability continually by means of non-formal and formal training. In an integrated and complementary manner, formal and non-formal training contribute to the training, molding and professional development of the individual for the library and information profession. (6).

The Aim of Training and Development

The aim of training is to promote greater efficiency of employees in their current posts and to develop employees with a view to additional or other responsibilities. Training and professional development also promote an increase in productivity.

Although the majority of library and information workers have already completed their academic education or are at least studying part-time when embarking on a professional career, further training on a regular basis is essential for several reasons. Orientation training is essential to familiarize employees with their new working conditions. This creates the opportunity to exploit an employee's productivity from the day of his appointment. In addition, regular and planned training prevent newcomers from getting mixed up by fragmented information. Intensive and purposeful training can shorten the "learn time" of workers. The skills and knowledge individuals acquire over a longer period of time by means of trial and error can be acquired in a more effective way with an intensive training programme. The idea of "throwing someone in at the deep end" does not benefit the organization.

Training is not only aimed at recruits in the profession. Further training is necessary because library and information services, in most cases, form part of public institutions with a commitment to the users to render a service of high standard at all times. This standard for effective service and the right attitude can be achieved and molded by training.

The objective of tertiary education is to train and prepare individuals in such a way that they can develop fully in their careers. Moral support and the recognition of training and further development by management is an aid to ensure the loyalty of employees and to keep their morale high. Training also assists in preparing workers for advancement. In library and information organizations, where there is a closed personnel system and where promotion is based on merit, planning training programmes is essential to offer equal opportunities to all employees.

The changing demands of the library and information profession dictate that preparation for this field cannot be regarded as completed. Professionals must take the changes in the world into account. To perform their tasks successfully, individuals must update their personal skills constantly by means of further training and development. Regular training and the involvement of employees in different types of professional development activities create the opportunities for employees to become more valuable to their library and information organizations.

Who Should Initiate Training?

Creating opportunities and facilities for professional growth is the shared responsibility of individuals, the profession, professional associations and library schools. For individual library and information workers, employee development is a lifelong process. The responsibility for this rests with the individual library and information worker. These persons can only develop if they are willing to do so. If individuals show the ability to develop further but are not interested in doing so, they can and should be positively motivated. A prerequisite for professional development is a positive attitude towards self-development and a sound self-image. Individuals must initiate the change, and they should be made aware of the opportunities to improve and refine their professional image.

It is not possible for library school faculty to graduate "tailormade" professional library and information workers. The library and information managers in practice have an obligation to provide effectively for staff development to orientate individuals for specific jobs. These managers should offer their employees the opportunity to receive continual education inside or outside the organization. Support by employers can take the form of granted leave, the financing of courses, and the recognition of regular and applicable continuing education when promoting employees.

Continual education is much more significant when it is pursued meaningfully and progressively. Employers are responsible for seeing that personnel participating in continuing education activities do so in a planned manner. Personnel participating at random in continuing education activities will be less productive for the organization. Employers can help plan with employees for attendance and possible follow-up activities after having discussed the needs of the employees for continuing education. Employers should also, particularly because of their pro-active approach to management, identify areas in which continuing education is needed. When employees' needs cannot be met, employers should take the initiative to present appropriate courses, possibly in cooperation with another employer. Training institutions or professional associations can also be helpful.

Library and information management identify new or adapted needs from the users and the community. These needs can be converted into services to users through expanded services. The expansion of services requires well-prepared staff to put services into practice. As stated previously, appropriate continuing education can compensate for possible shortcomings in the initial training of library and information workers. Through continuing education, library and information courses we can continue to contribute towards an effective and dynamic service.

It is possible for an employer to make continuing education activities available to persons in other organizations. This sharing can serve as a marketing forum for the organization. However, personnel time and costs involved should be evaluated in terms of the potential advantages to the organization. Sharing can also be cost effective. A considerable amount of money and manpower might be saved if employers who have similar library and information users. It is possible that a joint effort can be used to renew knowledge and skills. (7)

Professional associations can assist continuing education. In terms of the policy of the South African Library Association, applied from 1930 to 1980, the holiday schools arranged by the State Library played an important part in the development of libraries in South Africa. The holiday schools also satisfied the need for trained personnel in the first years.

Under the South African Institute for Library and Information Science (SAILIS), the Committee for Non-Formal Education exerts itself on behalf of all matters involving non-formal training. However, it is particularly the interest groups, functioning from the various branches, that are active with regard to training. SAILIS provides opportunities for individuals who want to become involved in further training.

Finally, responsibility for continuing education is also a part of tertiary training institutions. Training for professional library and information workers, undertaken by universities, prepares students for their occupation; however, the training offers only the essential theoretical background and limited experience. Yet, academics who are in touch with practical work situations can help identify any shortcomings in the earlier education of students and can prepare the necessary continuing education opportunities. Library school faculty at universities should:

- Keep practitioners informed of innovations in knowledge and technology, thus minimizing situations where graduates are faced with situations showing their lack of experience or preparedness.
- Develop lecturers as researchers to determine effective teaching methods to meet the needs of adult students.
- Prepare continuing education programmes based upon the new professional knowledge.
- Help evaluate continuing education programmes to see that they are up-to-date and meet the changing needs of the practitioner.

Because tertiary education institutions have bureaus for continuing education, personnel in these units can help library school faculty plan and organize continuing education experiences. Bureau personnel are also helpful in planning cooperative ventures with the government, the practical work situation, and other education institutions.

Analysis of Training Needs

The effective attainment of an organization's objectives should be the final goal in any training and development strategy. The organization's short-term and long-term goals should be analyzed. The planning, development, and maintenance of human resources in a library and information service organization should be effective. Training and staff development play a significant role in bringing together the skills of workers and the requirements of the organization. (8) The analysis of training needs in a library and information service organization takes place at three levels, organization, occupational, and individual.

By studying the objectives of the organization and the short-term and long-term plans to achieve these objectives, it can be determined whether new or changed training needs have emerged. If objectives of the organization are not achieved, new training programmes are probably needed.

At the occupational or operational level, job descriptions indicate the necessary skills and abilities of workers. If workers do not comply with these requirements or if their abilities are not sufficient, appropriate training can serve to bring them up to standard.

Work performance and efficiency of individuals can be measured according to predetermined standards. If they do not comply with these set standards, training needs are established. Employees who do not meet standards may exhibit the following:

- Low productivity that may be the result of insufficient, faulty, or poor training;
- Slow or poor service may indicate a need for training if the worker has been forced to use trial-and-error methods, therefore feeling insecure;
- Poor supervision or poor guidance by supervisors can result in poor performance by workers. Continuing education experiences for supervisors can help them become better managers of personnel;
- Training can be regarded as a means of actualizing the pursuit of success. Those people who have an inherent need of power can attain their desired goal through focused training.

Advantages of Training

Beach (9) explains the advantages of effective training for any organization as follows:

- By means of appropriate training, employees quickly learn to carry out their tasks according to the required standard.
- Training enables employees to increase their levels of success, which can lead to an increase in, and improvement of, output.
- Employees who are well-trained need less supervision.
- During training sessions, employees are exposed to the same inputs and are therefore more loyal to and involved in the organization.
- Employees who extend their knowledge and acquire skills enhance their own value in the labour market. It gives greater job satisfaction and creates the opportunity for promotion to more responsible positions.

Evaluation of Training

In order to determine whether training activities actually contribute to increased effectiveness or not, training programmes should be evaluated. Evaluation can take place at four levels as follows:

- The reaction level of course attenders after having completed the training session.
- The "learning level" which determines to what extent new knowledge has been gained.
- The behavior level of employees which should reflect a positive change after training.
- The result level, where it can be determined whether training in fact led to measurable results such as improved service or an increase in productivity.

Contemporary organizations cannot counter change. Employees should first take the initiative to develop themselves, but employers should also create the essential infrastructure and give employees their support so that they know that they can make progress in the organization.

Regular participation in training activities aimed at professional development integrate the objectives of the organization with those of the individual to the benefit of both parties. The employee will experience satisfaction and personal development and the organization will have the advantage of increased productivity and creativity.

References

1. *Inleiding tot Die Bestuurswese* geredigeer deur G.J. de J. Cronje, E.W. Neuland & M.J. van Reenen. Johannesburg: Southern. 1987, p. 391.
2. Megginson, L.C. *Personnel Management: A Human Resources Approach.* Homewood, Ill.: Irwin. 1981, p. 205.

3. Ivancevich, J.M. & Glueck, W.F. *Foundations of Personnel/Human Resource Management.* Plano, Tex.: Business Publications. 1983, p. 14.
4. Megginson, *op. cit.*,
5. Robbins, S.P. *The Administrative Process.* 2nd ed. Englewood Cliffs, N.J. Prentice-Hall. 1980, p. 11.
6. Terblanche, F. *Die Funksie van Voortgesette Opleiding by Verandering in die Biblioteek-en Inligtingswese in Suid-Afrika.* M.Bibl.-verhandeling. Pretoria: University of South Africa. 1988, pp. 70-72.
7. Webb, S.P. *Personal Development in Information Work.* London: ASLIB. 1986, pp. 138-141.
8. Gerber, P.D., Nel, P.S. & Van Dyk, P.S. *Mannekragbestuur.* Johannesburg: Southern. 1987, p. 207.
9. Beach, D.S. *Personnel: The Management of People at Work.* 5th ed. New York: Macmillan. 1985, p. 244.

Additional Reading Material

Andrews, Y. *Die Personeel Funksie.* Pretoria: HAUM, 1987.

Corbin, J. "The Education of Librarians in an Age of Information Technology." *Journal of Library Administration.* 9(4): 1988.

Richie, S. *Training and Management Development in Librarianship.* London: British Library Research and Development Department. 1988.

STEPS IN THE DESIGN AND DEVELOPMENT OF DISTANCE LEARNING COURSES

Laurel A. Clyde
Senior Lecturer
Department of Library and Information Studies
Western Australian College of Advanced Education,
P.O. Box 151, Nedlands Western Australia 6009, Australia

ABSTRACT

Distance education can be used to provide opportunities for continuing professional education. This article introduces the concept of distance education, suggests methods for identifying needs, provides steps for planning programmes, and describes programmes designed and implemented for continuing education through distance learning in Australia.

Introduction

Distance education, correspondence courses, off-campus study, external study, remote learning, extra-mural study, extension teaching, open learning, home study, are all terms used to describe a method of teaching and learning where the learners are normally at some distance from the institution or organization with which they are enrolled. Students study at home and do not attend classes on the campus or in the buildings of the institution or organization on a regular basis.

Learners are normally provided with a syllabus, a reading list, and some study materials. Depending on the institution or organization, these materials may be quite sophisticated and include both print and non-print materials. Learners may or may not be expected to spend some time at the institution or organization, e.g. "residential weekends," or "vacation schools".

Other opportunities for contact between teacher and learner may be provided, ranging from correspondence and telephone contact to teleconferencing, facsimile transmission, video conferencing, electronic mail, and electronic bulletin boards. Sometimes courses may actually be based on radio or television transmission or on computer communications, in which case, terms such as "telelearning" or "the electronic classroom" are used. Some courses assume the formation of local groups for discussion and interaction and provide guidance for such groups. There may or may not be a back-up library service for learning.

Distance education can be used to provide formal professional courses; librarianship courses at the degree and postgraduate level have been available in this way for many years, notably in countries like South Africa and Australia. In addition, distance education can be used to provide opportunities for continuing professional education.

As a vehicle for continuing professional education, distance education has many advantages. Usually it makes it possible for learners to undertake a course when they need to do so rather than waiting for a course to be organized in their own local area or having to travel some distance to attend a course in a remote city. Learners can work at their own pace, and skim, re-read, or revise as necessary for their own learning. Students are not constrained by the fastest or slowest learner in a class. The study materials are always available, whereas a lecture or workshop that is imperfectly understood cannot normally be repeated. Learners can study in whatever time they have available, whether that is late at night, early in the morning, or at weekends.

External study can be fitted around a normal job. Needs of people who are too remote from a large centre of population to have access to regular continuing education activities can be met through distance education as can those of people with very specialized needs. Distance education can also meet the needs of people in cities who are, nevertheless, unable, for one reason or another, to attend classes or workshops.

Distance education, in the context of continuing professional education, also has disadvantages and limitations. These can be minimized through good course design, but they do need to be taken into account by both teacher and learner if distance education is to be successful. Among the disadvantages and limitations, the one most often mentioned by participants in distance education courses is the lack of face-to-face contact between teacher and learner and between learners. It is clear that this personal contact is a factor in encouraging the learner to persist with a course and to open up to new ideas. To a certain extent, this can be overcome through structured telephone contact, but clearly "the isolation of the long distance learner" remains a problem with which course developers have to deal.

Other problems faced by the remote learner may include mail delays, the lack of local help when problems arise, e.g., equipment or software which is often identified as a lack of appropriate resources in small or isolated communities. While the latter problem can sometimes be overcome by supplying learners with everything they might conceivably need in order to complete the courses, this itself presents problems if the aim of a course, as it so often is in librarianship or information studies, is to help people to learn to locate the information themselves.

Steps in The Design and Development of Courses

In simple terms, success in providing distance education is dependent upon collecting information before planning begins. It is essential that the developer knows what is to be taught, knows who the learners are, and knows something about the learners' characteristics including their needs and expectations and their location to receive the programme. Knowing what resources, equipment, services, and advice or advisors will be available to the learners is essential; also, understanding which instructional strategies and media will best suit the subject content and the needs of the learners is necessary. Plans must be made for evaluating the results both in terms of summative evaluation of the student and formative evaluation of the distance education programme.

The above is, of course, a standard instructional model, familiar to all people who conduct courses on a regular basis in face-to-face settings. However, it is surprising how often this model is ignored in distance education, as though, in moving to a new mode of instruction, people forget the validity and effectiveness of established models. Yet, this one translates well to the distance education setting, though the actual processes may be carried out in very different ways when planning and developing distance education courses.

Identification of the need for CPE programmes by distance education

It is axiomatic that, if they are to be successful, distance education programmes, as other continuing professional education programmes, must meet a need. The accurate identification of areas of need is crucial. In relation to distance education, there are two aspects: the identification of topics and developments which are likely to interest practitioners or which will be important for them in the workplace; and the identification, from among these topics and developments, of those which can appropriately be addressed through provision of distance education programmes.

Generally speaking, the development of distance education programmes is appropriate when a topic is important for a large group of the profession, whether or not they are close to a large centre of population where courses and seminars might be provided, e.g. information skills in schools. A topic or development might be particularly related to rural libraries or to people who are isolated for whatever reason. Additionally, a topic might be important for a specialized group whose membership is small and widely scattered, e.g., map librarians or sound archives managers.

Research must play a large part in the identification of the need for CPE programmes by distance education. Practitioners who do not get the opportunity to attend conferences or read the current literature may be unaware of their need, simply because they may be unaware of likely future developments or even of current developments. They may not always be in a position to define their needs to course providers.

In this research, the professional associations and the library schools have a role. Sometimes, a need will become very obvious to the members of professional associations or to employing authorities, e.g.,when a new educational policy directive has implications for library and resource services in schools or when a change of government policy or new legislation has implications for public libraries. Changes in copyright legislation in many countries have had implications for all types of libraries and have led to a need for continuing education opportunities. The same is true of some developments in technology.

Defining Potential Markets: Market Research. Many people see "market research" only as synonymous with selling and promotion in terms of presenting a product. However, market research also encompasses needs assessment, product development, marketing research, pricing, and distribution. All these are valid activities in terms of CPE course development.

> Traditionally, it is the business community that has carried out marketing research. The educational sector has largely stayed out of it. However, the spiralling costs and restricted budgets are forcing educational institutions [in the widest sense] to take a closer look at their target markets. There is thus a growing need to link marketing as a useful tool for a non-profit institution, such as a university, as opposed to a commercial organization. (1)

While a need for continuing education may be apparent to the organizations and institutions who are likely to provide appropriate courses, not all practitioners are likely to be convinced of the need to enroll in the course. Further, some may already have had an opportunity to acquire the information or skills. Thus, the potential market will be only a proportion of the group identified as needing the course. Often it is necessary to develop a "prospective participant" list.

Directories of libraries and information agencies or the mailing lists of professional associations may be used to identify individuals or groups with a potential interest. Advertising or articles in professional journals and newsletters, in trade journals, in trade union newspapers, in major state or national newspapers, and on online information services may be ways of targeting particular groups.

At Western Australia College of Advanced Education, where courses are offered for two district groups, teacher librarians and library technicians, different outlets are used to target the different markets. To reach teacher librarians working in schools, the teachers' union newsletters are used a great deal, since these newsletters reach people in charge of school libraries whether or not they are members of a professional association.

To reach library technicians, direct mailings to employing libraries, based on the various libraries' directories, are the principal means used. These techniques were not arrived at by accident but through the results of questionnaire surveys of course participants. The findings of the surveys indicated where participants first heard about the course in which they had enrolled. These methods were considered the most successful vehicles for each of the different groups. In addition, surveys of practicing teacher librarians and library technicians were conducted to identify journals, newsletters, and newspapers which were most frequently read by each group so that these could be used for publicity and promotion.

Sampling techniques will give an indication of the number of people in the potential market who are likely to enroll in the course. Even so, the results should be treated with caution, as our experience of matching the survey results with actual subsequent attendance, shows that not all of those who indicate an interest in the planned topic or course can be expected to enroll. Personal factors, cost, time of year, a change of job, can all have an influence on enrollment patterns.

Identification of learner characteristics

Before a CPE distance education course can be planned, the organizers need to know something about the specific needs of the potential learners in relation to the topic, their prior knowledge and skills, their work situation, the resources such as facilities and support available to them locally, and their prior experience of the type of continuing education activity envisaged. Because the course participants are likely to be spread over a wide geographical area, course developers also need to know if some groups of potential distance students are likely to be more disadvantaged than others in terms of access to libraries, resources, media facilities, or whatever is needed to complete the course adequately.

Once again, survey and sampling techniques can be used, with the sample population derived from those identified as being in the potential market group. This step can be a critical one in the design of a successful course, and, particularly in identifying possible areas of difficulty. For instance, at the Western Australian College of Advanced Education, a questionnaire survey of potential enrollees in a distance education course on library automation for library technicians showed that most of the group were living in the country and working in small country libraries. Yet, some 73.53 percent were working in libraries where computers were being used for some tasks.

The percentage was a much greater proportion than had been anticipated, given information from other sources. The course developers realized that the course as a whole should be prepared at a higher level than had been originally intended. The same survey identified a possible source of problems. Although the figure of 73.53 percent applied across Australia, in some Australian states, the figure was much higher; in others, it was much lower. For instance, 92 percent of potential external students in Queensland were working in libraries where computers were being used, but the figures were only 16.66 percent for Western Australia. (2) Clearly, some allowance would need to be made for this in course development.

Selecting appropriate content and skills to be taught

The selection of specific content and skills to be taught through the CPE distance education course is very important if needs are to be met. It is partly dependent on a knowledge of the needs of likely enrollees and partly on a knowledge of the workplace and its needs. The former should become apparent through identifying learner characteristics. The latter will require input from those managing libraries and information services.

This input may come through surveys. It may also come through having representatives from the practitioner groups on the organizing committee. Where the practitioners work in remote locations, their input may need to be sought through teleconference meetings. Once again, telephone or questionnaire surveys have their place. In some cases, such as where a course is planned in response to a previously-identified need, this information may already be available.

Selecting appropriate learning media

Traditionally, distance education has relied on print materials, reflecting its nineteenth century origins in correspondence schools. Later, audiotapes, color slides, slide and tape sequences, simulations, games, videotapes, were added to the instructional media from which course developers could choose.

Kits of samples or realia have also been popular in distance courses including such things as packages of bookbinding materials, photography chemicals, or sets of catalogue cards to be filed in sequence.While it is a little outside our field, one Australian distance education course on wine appreciation even includes a substantial kit of miniature bottles of wine, with tasting notes!

New technology has opened up new possibilities for distance education; media currently in use or being tested for distance education include floppy disks, videodiscs and CD-ROM, microfilm,and also broadcast and narrowcast television via satellite and cable television. Audio teleconferencing and video conferencing have also been popular as interactive media, as have the use of online information services, electronic bulletin boards, and electronic mail.

Given the relatively high degree of access which library workers have to computing facilities and electronic mail, it is likely that some form of computer communications will play an important part in distance education CPE programmes in the future. Computer communications could possibly deliver some course materials, but it is more likely to be used in the area of communication between teacher and learner and between learners in different locations.

Normally, in any distance education course, a range of instructional media will be used in combination. Some will be used for delivery of instructional content, some to provide enrichment materials, some for communication and feedback, and some for all of these. Choice of media will depend on the nature of the subject matter and skills to be covered, the knowledge level and prior skills of the learners, and the learners' access to replay equipment. Access to television sets, video playback equipment, microfilm readers, and computers must be available if used in course development.

Finally, the choice of instructional media further depends upon the skills and budget of course developers. Persons who do not understand instructional media will not understand how to develop these for distance learning. Development of excellent instructional media requires an initial outlay of funds for the creation of the products. Many course developers have limited budgets.

Designing print materials to maximize learning

In 1981, Marland and Store reviewed the literature and practice of distance education as the basis for an analysis of the components which make for effective distance learning materials. They noted then "...a preference for one particular approach to teaching through text" (3). Though the last ten years have seen an increasing reliance on other forms of course delivery media, most distance education courses, even those based on teleconferencing, use printed materials to some extent. Store and Marland noted that print materials usually incorporate devices:

> which have been derived from, or can be readily aligned with, a traditional approach to teaching in which, typically, specific goals are defined for the students; content and resources are presented by the teacher; interactive processes and patterns are largely initiated, guided, and controlled by the teacher; obligatory assessment activities are set by the teacher; and feedback is provided by the teacher. (4)

The reasons for this teacher-directed approach are complex and often educationally valid. Further, they give the instructor a sense of control over the student despite the distance between them. However, research quoted by Marland and Store indicates that this may be an illusion.

In support of this, a study carried out in 1981 at Townsville College of Advanced Education (5) suggests that, regardless of the attempts of the designers of external study materials to guide distant students through a pattern of reading, activity, review and consolidation, remote students will approach the study materials in the light of their own needs. Diagram 1 on the opposite page shows, for instance, how six different external students in this research study approached the same set of learning materials (module 3 of a Mathematics in-service course for teachers.) The course was based on very structured print materials. The fine lines indicate in-depth, slow reading and activity, and the broken lines indicate back-tracking through the study materials. The heavy dots show the beginning and end of a study session.

From the literature and from evidence such as this, Marland and Store concluded that what distance education students need in print materials is an "access structure" so that they can find their way around the text and locate and use the material that is most relevant to their needs. They went on to indicate the design features of instructional text which would facilitate learning for remote students. These features include the following:

- Devices which orient the learner and introduce the material to be mastered such as advance organizers, overviews, pretests, stated objectives;
- The insertion of questions in the text, the type of question, purpose, frequency of questions, and model answers;
  and,
- Print presentation techniques which have implications for the effectiveness of instruction, including booklet layout, page layout, graphics, and typographical cues such as type size, type style, color, and emphasis including heavy type and use of type in the margins.

Designing non-print materials

Much has already been written about the design of non-print instructional materials, particularly in the literature of educational technology. Books such as Brown, Lewis, and Harcleroad's *AV Instruction: Technology, Media, and Methods*, Kemp's *Planning and Producing Audiovisual Materials*, and Gerlach and Ely's *Teaching and Media: A Systematic Approach*, will help.

Articles in journals such as *Educational Technology, Media and Methods*, *The Australian Journal of Educational Technology*, *The Computing Teacher* will be useful. Each medium has its own requirements which must be taken into account in the design of study materials. Once again, the subject matter to be taught and the needs of the remote learners should be kept in mind in the materials design process.

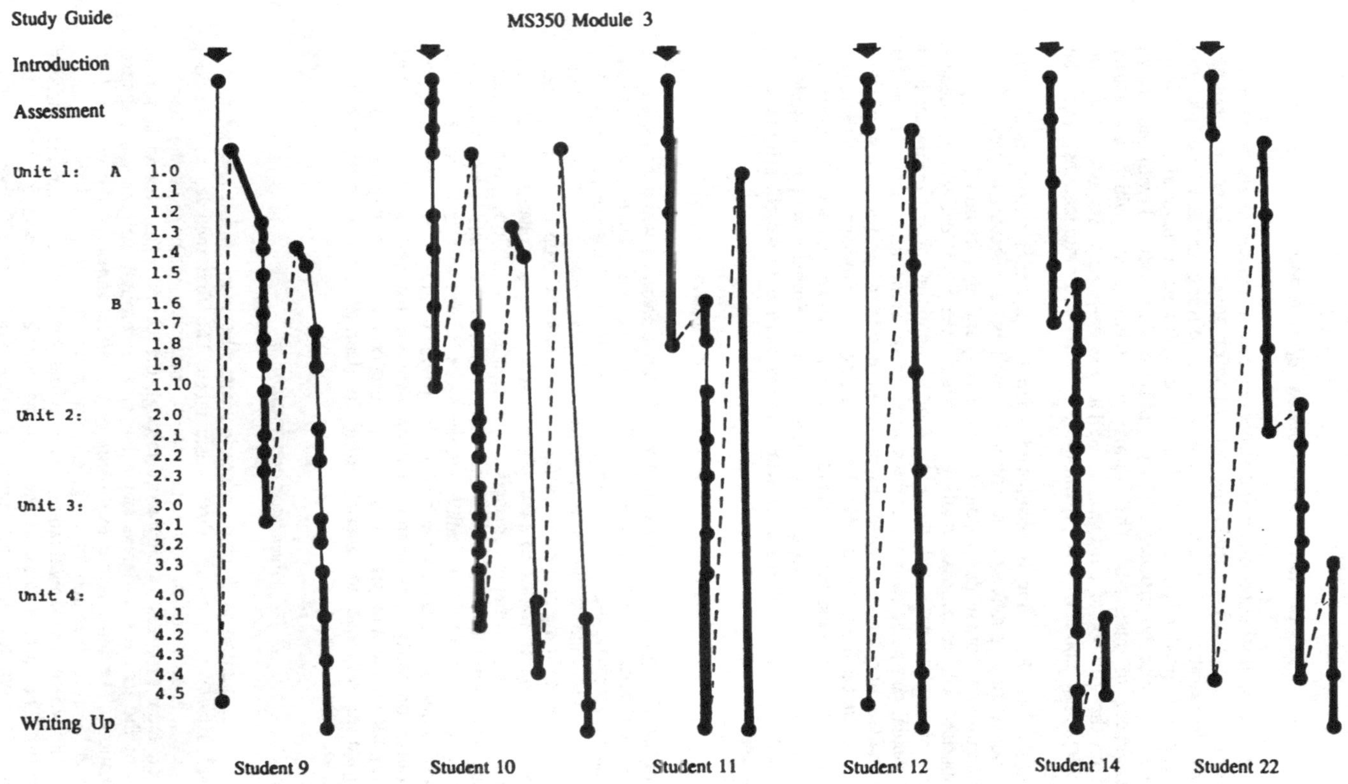

MS350 Module 3
Study Guide
Introduction
Assessment
Unit 1: A 1.0
1.1
1.2
1.3
1.4
1.5
B 1.6
1.7
1.8
1.9
1.10
Unit 2: 2.0
2.1
2.2
2.3
Unit 3: 3.0
3.1
3.2
3.3
Unit 4: 4.0
4.1
4.2
4.3
4.4
4.5
Writing Up
Student 9
Student 10
Student 11
Student 12
Student 14
Student 22

Preparation and distribution of materials

Any organization or institution offering distance education courses needs mechanisms and procedures for preparing, producing, promoting or advertising, and distributing the courses. Preparation of materials will include printing of text and pictorial materials and the production and duplication of audiovisual materials. Broadcast programmes must also be produced. There will be costs associated with the design and development of materials. The normal production costs, which are partly dependent on the number of enrollees, as well as the on-going costs associated with mailing, telephone contact, and electronic or other communications must be considered.

The basis for advertising is established when the potential market is defined. Costs for advertising, publicity, and promotion will be a substantial part of the budget structure for course development, production, and distribution.

Distribution will encompass mailing or other means of delivery of course material to distant students and the use of other communications media such as broadcast media, computer communications, and teleconferencing. If course materials are to be mailed, consideration needs to be given to this in the production process. Booklets need to be light, yet sturdy; the packaging of audiovisual materials needs to be designed to withstand the vicissitudes of postal services. Sometimes, there are postal regulations which need to be taken into account. Sometimes it is possible to get reduced postal rates for material which meets particular weight criteria or is packed in certain ways.

Communication with distance learners

Possible means of ongoing communication with distance learners while they complete the course have already been mentioned. They include mail, telephone, facsimile, teleconference, electronic mail, interactive video, even the exchange of audiocassettes. Some form of communication is a feature of many, though not all, distance education courses. Communication is useful to help students keep on task, to enable discussion about course topics to take place, and to enable students to test their understanding of subject matter or their acquisition of skills. Good communication can reduce the drop-out rate and can assist students in their preparation for examinations or to meet accreditation requirements where this is a factor in the course. Sometimes additional course content is provided through this communication as the course proceeds. The communication between teacher or institution and learner may be formal and planned as part of the course, or it may be unplanned and informal in response to queries from the learner through mail, facsimile, telephone, or other means.

Feedback and evaluation

Feedback and evaluation in distance education can take many forms and be formal or informal. However, there are three basic types. The first includes self-checking exercises, tests, and activities within the course materials which give the learners feedback on their progress as they work through the course. The second is formal evaluation of the learners' progress and achievements through assignments and texts or examinations. The third is the evaluation of the effectiveness of the course itself in achieving its objectives and meeting learners' needs.

The first two forms of evaluation relate to the needs of the learners enrolled in the course. The third relates to the need of course developers for information which will help them to adapt this course or to develop more effective courses in the future. All three forms of evaluation can be achieved in many different ways. What is crucial is that techniques of providing feedback to students on their progress and assessment and evaluation strategies should be an integral part of the course. Strategies should be planned as part of the course from the beginning of the course development process, and they should be related to the needs of learners and course developers. Finally, these strategies should be appropriate to the subject matter or skills which form the focus of the course.

Distance Education Courses; Some Australian Examples

While pre-professional education for library and information work in Australia has long been characterized by a strong emphasis on distance education, it is only in the last 10 to 15 years that distance education techniques have made any impact on continuing professional and paraprofessional education in this field. However, the last decade in particular has seen the implementation of a number of such projects at the national, state, or local level.

Some of these projects are large, technologically based, and well-funded. Others are small and developed "on the cheap" by concerned professionals who have identified a need and attempted to meet it. Among the former are the continuing education programmes on information technology offered via radio, with supporting materials, by the University of New South Wales in Sydney and the TeleSLAQ series of state and national teleconferences for teacher librarians, organized by the School Library Association of Queensland (SLAQ) in association with Brisbane College of Advanced Education.

Among the smaller projects are the *Priority One: Information Skills* continuing education kits prepared by the Australian Library and Information Association's School Libraries Section (NSW) Group and the *Computer Awareness for Teacher Librarians* kit developed by Brisbane College of Advanced Education. Three of these projects will be described in more detail so that readers have some idea of the potential of distance education.

TeleSLAQ, a continuing education programme for teacher librarians, is one of the larger projects. Originating in Queensland, a large state with many isolated schools and libraries, this programme aims to bring quality in service and continuing education to teacher librarians within their own local areas or within reasonable traveling distance.

TeleSLAQ allows people who live a long way away from capital cities to discuss issues in school librarianship with experts in Australia and overseas through telephone links. It also helps to overcome problems of professional isolation by bringing together local groups, however small, to take part in state or national teleconferences.

The first four TeleSLAQ teleconferences took place in 1983. Because of the success of these original teleconferences, the idea was redefined and enhanced over the years, and 1989 saw four TeleSLAQ V programmes being presented. While audio teleconferencing has been the basis of TeleSLAQ, a range of other technologies and communications media have been used including video conferencing, videotape, audiocassette, computer links, and print media.

People in between 10 and 20 centres take part in each teleconference. Participants are supplied with a pre-teleconference booklet giving instructions for the teleconference, technical details, and pre-teleconference readings. Sometimes a videotape, audiotape, or computer programme is a part of the pre-conference mailing.

A local leader convenes each group before the teleconference to discuss the issues, view any audio or video tapes, and prepare discussion questions. The actual teleconference link-up provides the basis for an exchange of ideas between all participants in all centres and gives participants the opportunity to ask questions of an expert panel in Brisbane.

After the link-up, the local groups hold further discussion on issues raised during the teleconference. This provides an opportunity for interaction at the local level and for preparation of formal feedback to the Brisbane organizing group.

Local groups may range in size from two or three people to 30 or so. Altogether, 200 to 300 people may take part in any one teleconference. The materials prepared for the teleconference programmes can also be used by isolated individuals. Topics which have been explored in TeleSLAQ teleconferences include "Audiovisual Applications of the Microcomputer," "Cooperative Networking as a Basis for Resource Sharing," "Using Non-Fiction in Years 11 to 12," "Storytelling Across the Curriculum," and "Cooperative Planning and Teaching"."

Although the TeleSLAQ concept places demands on the course organizers in Brisbane and the local leaders and requires considerable effort in preparation of resources and teleconference booklets, its success has been a great encouragement to all involved. Some 20 TeleSLAQ teleconference programmes have been organized to the end of 1989 with large numbers participating in each, many of whom are working in very remote areas. The opportunities for interaction have been appreciated, and the "spin-offs", the print and non-print materials created for the teleconferences, have reached other people who could not take part in the teleconference discussions.

One of the smaller projects, a kit, was developed in 1987 when the "Information Skills Across the Curriculum" concept was introduced in New South Wales. The School Libraries Section (NSW Group) of the Australian Library and Information Association prepared a continuing education kit, *Priority One: Information Skills*.

This kit was designed to be the basis of a series of workshops which could be run as a full-day seminar for teachers and teacher librarians or as a sequence of shorter workshops conducted over a period of time. Equipped with the kit, a leader in a region or in a school could conduct the workshops.

The kit provided a suggested programme, notes for the leader, activities, an audiotaped talk from Joyce Kirk, an internationally-known expert on information skills, a set of overhead projection transparency masters, discussion materials, session notes, and resource materials. It thus provided a means for local leaders, however unskilled, to bring a well-supported seminar to people outside the metropolitan areas. The fact that more than one thousand copies of this kit were sold in a two-year period suggests that individual teacher librarians also used it as a means of personal professional development; in fact, as a distance education project.

Another of the smaller projects, *Computer Awareness for Teacher Librarians* kit, was developed in 1981, as an individual, self-instructional kit. At this time many Australian school librarians were beginning to consider library automation seriously, yet very few teacher librarians had any computer skills. The course materials, prepared by Peter J. Pegg for distribution through the Department of School Librarianship at Brisbane College of Advanced Education, comprised a set of study notes, readings, and three floppy discs for the Apple II computer that was then, and for some years, the most popular computer in Australian schools.

The aim of "this self-instructional continuing education course [was] to enable participants to become aware of the potential and the limitations of microcomputers and to gain some basic skills". Exercises were "orientated towards school library applications".

The kit "stood alone". A feature of the kit was that it incorporated, on disk, a set of "real" library programmes which formed the basis of some of the exercises. Topics covered included basic computer concepts, operating a microcomputer, and applications software, including library software, e.g., overdues programme and a programme to produce lists of videotape holdings.

Self-correcting review questions and computer-based exercises were provided to assist learning, but no feedback to learners was provided. If people reached the end of the kit, then they had successfully completed the course. However, a course evaluation form was provided for comments which users could return to the College. This helped course development but not the learners.

These three examples represent very different, yet valid approaches to continuing professional education for people in country or remote areas. Two of the three, while being based on distance education techniques, incorporate a means for bringing small groups of people together within a local area in an effort to overcome the problems of professional isolation.

This paper has barely scratched the surface of a complex area in which there is a rapidly-developing literature. Further information is available through national and international associations in this field, such as the International Council for Distance Education (ICDE), the Australian and South Pacific External Studies Association (ASPESA), and the Canadian Association for Distance Education (CADE).Conferences on distance education are held by these organizations, and specialist national and international journals and newsletters are published. Among the better-known journals are *Distance Education*, the international journal published by ASPESA; *Teaching at a Distance*, from the Open University in the United Kingdom; and *Journal of Distance Education (Revue de l'enseignementa Distance)* (CADE). There are also specialist courses available for people involved in distance teaching, usually offered by distance education, too. Among these are the Graduate Diploma in Distance Education available internationally from the South Australian College of Advanceed Education, and similar courses elsewhere, including Canada.

References

1. Curtis, Frederick A. and Trilochan S. Bakshi. "Market Research for a Proposed Natural Resources Planning and Management Programme by Home Study." *Distance Education* 5 (September, 1984) p. 94.
2. Clyde, Laurel A. "Survey of External Student Access to Computers," *LiSt* (WACAE Department of Library Science Newsletter), August, 1989, p. 9.
3. Marland, P.W. and R.E. Store. *Some Instructional Strategies for Improved Learning from Distance Teaching Materials*. Townsville College of Advanced Education, 1981. p. 1.
4. Ibid.
5. Clyde, Laurel A. *et al.* "How Students Use Distance Teaching Materials: An Institutional Study." *Distance Education* 4 (1983) pp. 4-26.

Bibliography

Armstrong, J.D. and Store, R.E., eds. *Evaluation in Distance Teaching: Proceedings of a Workshop Held at Townsville College of Advanced Education in Cooperation with the Australian and South Pacific External Studies Association.* Townsville College of Advanced Education, Townsville.

Baron, Daniel D. "Distance Education: Removing Barriers to Knowledge."*School Library Journal* Vol. (November, 1989). pp. 28-33.

Bond, Sally. "Telecommunications Based Distance Learning: A Guide to Educators." Southeastern Educational Improvement Lab, Research Triangle, North Carolina. ERIC document ED 287 474.

Clyde, Laurel A. "Survey of External Student Access to Computers," *LiSt* (WACAE Department of Library Science Newsletter), August, 1989, p. 9.

Clyde, Laurel A. "Library Studies Courses by External Study: Background Paper for the Continuing Professional Education Round Table Workshop at the IFLA General Conference, Sydney." 1988.

Clyde, Laurel A. *et al.* "How Students Use Distance Teaching Materials: An Institutional Study." *Distance Education 4* (1983) pp. 4-26.

Curtis, Frederick A. and Bakshi, Trilochan S. "Market Research for a Proposed Natural Resources Planning and Management Programme by Home Study." *Distance Education* 5 (September, 1984).

Dawson, Merran. *Priority One: Information Skills*, School Libraries Section (NWS Group), Library Association of Australia. kit.

Gawith, Gwen. "TeleSLAQ II: A Teleconference on Library Networking." *New Zealand Libraries* 44 (September, 1985). pp. 207-209.

Hartley, J. *Designing Instruction Text.* London: Kogan Page, 1978.

Hough, Michael. "Motivation of Adults: Implications of Adult Learning Theories for Distance Education." *Distance Education* 5 (1988). pp. 7-23.

Keegan, D.J. *The Foundations of Distance Education*. London: Croom Helm, 1986.

Kemeny, Linley. "Teleconferencing: an Interactive Medium." *Education Library Service Bulletin* 22 (Second Term, 1984). pp. 5-6.

Marland, P.W. and Stone, R.E. *Some Instructional Strategies for Improved Learning from Distance Teaching Materials.* Townsville: Townsville College of Advanced Education, 1981.

Mason, Robin and Anthony Kay, eds. *Mindweave: Communication, Computers, and Distance Education.* Oxford: Pergamon, 1989.

New South Wales. Department of Education. *Keylink Users' Guide*. Sydney: Computer Education Unit, Department of Education, 1989.

Pegg, Peter J. *Computer Awareness for Teacher Librarians*, Brisbane: Department of School Librarianship, Brisbane College of Advanced Education. (kit).

Rowntree, Derek. *Teaching Through Self-Instruction: A Practical Handbook for Course Developers.* London: Kogan Page, 1986.

Rumble, G. *The Planning and Management of Distance Education.* London: Croom Helm, 1986.

Stewart, D. and J. Daniel, eds. *Developing Distance Education.* Oslo: International Council for Distance Education, 1988.

Thomas, R. "Online Adult Education in the Light of the Open University Experience," In: *Online Information 89, Proceedings of the 13th International Online Information Meeting, 12-14 December 1989, London, England.* London: Learned Information, 1989. pp. 303-319.

Thorpe, Mary. *Evaluating Open and Distance Learning*. Essex, Longman Open Learning, 1988.

Western Australia. Department of Education. Audio-Visual Education Branch. *Teleconferencing in Education*. Perth: Audio-Visual Education Branch, 1984.

Western Australian Satellite Communications Eduation Advisory Group. *Educational Broadcasting and the National Communications Satellite*. Perth: WAIT, 1981.

TUTORIALS: A NETWORK STRATEGY FOR LEARNING

M. Kent Mayfield
Executive Director, Education Foundation; and,
President-Elect
Continuing Library Education Network and Exchange
American Library Association, Chicago, Illinois, USA

ABSTRACT

Tutorials, an informal programme of free individual counselling on continuing education topics, provide an opportunity to review projects, discuss plans, try out ideas, and receive advice. This paper describes the process developed by CLENE to provide tutorials for members.

The last 15 years in the study of learning by adults has been characterized by research on self-directed learning and by increased popularization of the concept. Research has helped to dispel the notion that only learning that is institutionally arranged can be purposeful and effective and that self-directed learning in an informal setting is necessarily of a lower order.

Self-directed learning has been studied for a variety of learning populations. The picture that emerges from the research suggests that adults are choosing to plan, execute and evaluate their own learning activities and that many of them view this as the "natural" way to learn. Clearly this has practical implications for planners of continuing professional education for librarians.

One simple and practical strategy, exploiting the power of networking within the library profession, has been identified by the Continuing Library Education Network and Exchange Round Table (CLENERT). CLENE itself began as a collaborative initiative of several U.S. library associations, growing out of efforts to develop a national clearinghouse for continuing library education. It continues as a round table of the American Library Association. One of its primary goals is to foster networking and sharing of information among continuing library education providers.

For several years, beginning in the early 1980s as an activity planned around the midwinter meetings of the American Library Association and later extended to include the annual meeting, "tutorials" have been offered by CLENE. The tutorials are an informal programme of free individual counseling on library staff development/-continuing education topics. Generally limited to a period of one hour, each tutorial is tailored to the needs of a given situation.

For many participants, tutorials offer an opportunity to review projects, discuss proposed plans, try out ideas or receive advice, commentary and critique. In other cases, the tutorial is conceived as a more concrete training activity in a given area of continuing education. Examples of topics covered include statewide continuing education planning, workshop preparation, cross-training of staff for youth and public services, training for specialized and underserved populations, needs assessment, questionnaire development, marketing, and train-the-trainer strategies.

Tutorials are announced well in advance of the meetings for which they are scheduled, utilizing press releases, brief notices in library publications, including the *American Libraries, CLENExchange,* and *Library Hotline.* Tutorials may be requested by mail or by telephone, providing a brief description of a project or topic on which tutorial assistance is desired. In follow-up discussions with the CLENE Tutorial Co-ordinator, additional information is gathered, specific needs are identified and the scope of the tutorial is clarified. The coordinator then identifies a person qualified by experience and expertise to conduct the tutorial and willing to provide the counseling/-training at no cost. Thereafter, the selected tutor works directly with the tutorial participant to arrange a time, date, and place for the meeting. (To simplify the arrangements, CLENE regularly provides a suite of rooms in a conference hotel which may be used on a pre-established schedule for the tutorials.)

"Tutors" for the sessions are experts drawn from the CLENE membership which includes continuing education and staff development specialists in the U.S. and Canada. Tutors are selected to represent unique institutional settings, expertise and documented experience of specific relevance to the person requesting a tutorial.

Those requesting tutorials are often new to the field of continuing education or have acquired new responsibilities, such as the appointment of chair of a library's staff development committee, or as programme chair for a state professional association. Experienced practitioners also apply, however, to review existing projects, curriculum innovations or to extend their understanding of emerging, creative techniques in the field.

Critical to the tutorial process is the Tutorials Coordinator. Coordinators are responsible for arranging publicity for the tutorials, receiving and processing tutorial requests, identifying and contacting potential tutors to confirm suitability, availability, and willingness to participate. They are responsible, as well, for conducting a follow-up evaluation of the sessions, with reports to the CLENE Board of Directors.

Coordinators must be experienced professionals in library staff development and continuing education with substantial experience in the planning, delivery, coordination and evaluation of programmes. A wide network of acquaintances in both library/information science and adult education and human resource development is essential. They must possess excellent interviewing and interpersonal skills, in order to determine the needs to be addressed in the tutorial and to select a tutor appropriate to the situation.

Tutorials are one agency for information and skill exchange. They suggest ways by which adults can teach, stimulate, prompt, support, and learn from one another in a collaborative atmosphere. They can be managed at low cost with minimum staffing. As such, tutorials are a strategy which merits the attention of library educators and staff development specialists.

Additional information on the CLENE tutorials is available from CLENERT, American Library Association, 50 East Huron Street, Chicago, Illinois, 60611, USA.

POSTER SESSIONS AS A CONTINUING PROFESSIONAL EDUCATION TECHNIQUE

Katherine Harig
ALA Poster Session Committee Chair
9621 Pastora Place, Columbia MD 21045, USA

ABSTRACT

Poster sessions are an innovative technique for continuing education. They present the cutting edge of research and ideas in the field of library and information science and allow persons who have never participated actively at a conference to take part. This paper provides a background to the process and makes suggestions for the effective implementation of this CPE opportunity.

Poster sessions are graphic representations of an idea or concept; a solution to a problem; a description of research, or of an innovative programme. They are usually presented at conferences or conventions as an alternative or as a supplement to regular programmes and meetings, A Conference-Within-A-Conference.

Originally developed and presented by the medical and banking associations, these sessions since 1982 have been a regular feature of the American Library Association's annual conferences. They provide exhibit areas where juried graphic presentations are displayed for a specified length of time. No formal, verbal presentation is made. Instead, the essence of the ideas and concepts are conveyed using graphs, diagrams, pictures, and narrative text displayed on a four-by-six foot poster board.

Each presenter is given an exhibit space, usually ten foot square or greater so that those wishing to view the session can circulate freely and spend as much time as they choose per session. The presenter(s) of the poster session is on hand during the assigned period to answer questions informally and to further elaborate on the session.

A table is usually provided for handouts relating to the sessions, sign-up sheets for more information, and business cards of the presenter(s) for those who may not have the opportunity to speak personally with the presenter(s). Visuals such as photographs are often used to enhance the presentation.

In 1980, Dr. Elizabeth Stone, soon to become President of the American Library Association, was researching innovative techniques in continuing education. In studying other associations and their conferences, she found that many associations used juried poster sessions to present new research about their professions. Stone and others viewed poster sessions at the meetings of the American Chemical Society and the American Society of Microbiology, among others. A proposal to pilot-test poster sessions at an ALA conference was presented by Stone in September, 1980.

Poster sessions were presented as a method of streamlining conference programmes and a method to provide an opportunity for more individuals to participate in annual meetings. Poster session advocates state the following reasons for using poster sessions in continuing education. Poster sessions provide:

- More time to present more data.
- More visibility for more views for longer periods of time.
- Better two-way exchanges of ideas. (1)

Stone also noted the subsequent reuse of the materials "for teaching purposes or other functions."

Following "Stone's proposal, and in response to a need for change in conference format expressed in a 1975 survey conducted at the San Francisco ALA conference, an ALA committee was formed in 1981 to discuss poster sessions as a possible feature of future ALA conferences. The committee was composed of librarians from each of the ALA divisions and many of its sections. It was felt that such a broad-based committee would assure representation of poster session submissions from all types of librarians and libraries.

The 1982 Poster Session Committee included Dr. Fred Peterson as Committee Chair, Dr. Ruth Person as Review Subcommittee Chair, and Katherine Harig as Scheduling Manager. It also included 31 additional librarians to serve as reviewers of the abstracts required for submission.

At the 1982 Denver ALA Midwinter conference, it was decided that there would be seven broad groupings of topics. Potential presenters were provided with applications in an issue of *American Libraries* which included a sample abstract, a blank abstract form, and a list of topics. Applicants were asked to fill out the abstract form and choose which topic best suited their idea.

Abstracts were sent to ALA to be reviewed by the Review Panel. Accepted abstracts were sent to the Scheduling Manager who assigned slots of one-and-one-half hours based on no-conflict times during the conference. The seven broad session headings in 1982 included topics in all aspects of librarianship including public awareness, social issues, reference, technology, management, library education, literacy, and service to special groups. It was the intention that each broad session, such as reference, would include librarians from every type of library. Children's and school librarians, reference librarians, technical librarians, and academic librarians might intermingle to share ideas and concepts.

The first annual poster sessions were presented at ALA during its Philadelphia conference held from July 10-13, 1982. "Star-Studded Ideas for Libraries" included 94 poster sessions presented by 138 presenters. Four librarians presented two sessions each.

An abstract booklet composed of camera-ready abstracts of each session and an index were provided for attendees at each session. Attendance at the seven sessions was brisk and enthusiastic. The Public Relations section of the Library Administration and Management Division of ALA took slides of each of the 94 sessions and compiled a helpful slide-tape presentation to use for promotion, instructional, and archival purposes. The slide-tape has been duplicated and can be loaned on interlibrary loan from ALA's Headquarters Library.

It was decided that poster sessions were valuable enough to become an annual feature at ALA conferences. Approximately 655 sessions have been presented by 880 persons from 45 states and three countries since 1982. Many librarians have presented more than one poster session. One hundred and six have provided two or more, 33 have given three, 14 have presented four, 12 librarians have developed five poster sessions, and four librarians have presented six poster sessions since 1982.

Since 1982, poster sessions have become a vital part of conferences and meetings of such library groups as the American Society for Information Science, the Association of College and Research Libraries, the Library Information Technology Association, and many state and regional library associations. Conference planners refer to the ALA slide-tape presentation and consult with members of the ALA Poster Session Committee for advice on planning poster sessions at their conferences and meetings.

The tenth annual ALA poster sessions will be held with great fanfare at the 1991 ALA Atlanta conference. The popularity of the concept at ALA has required that available slots at the conference be increased to its present number of 108 sessions given over a three-day period. Approximately 2,000 conferees attend poster sessions per conference.

Poster sessions present the cutting edge of research and ideas in the field of library and information science. Conference attendees seek them out and make them an integral part of the conference agenda. Attendees enthusiastically report, "I never miss them [poster sessions]," "They are my favorite part of the conference," "You can manage your time and see what you want to see," "I learn more at poster sessions than at all the other meetings combined," and "I get to meet librarians who are doing exactly what I am doing. The contacts are extremely valuable." (2)

Advantages of Poster Sessions

Initially, poster sessions were thought to be a way to streamline conferences where attendees often missed meetings and felt that their time at large meetings was wasted. The ability for individuals to select the specific topic of interest and spend as much time as they choose viewing the sessions was an advantage often stated by attendees. Another advantage was the grouping of similar sessions together which provided the viewer with the opportunity to see the topic represented from the viewpoint of the entire profession.

Poster sessions are now scheduled at ALA during lunch hours, a no-conflict period. They are scheduled in the exhibit area where publishers and jobbers welcome the healthy influx of browsing librarians as they circulate through the aisles of the exhibits. Poster sessions are also a perfect vehicle for a small committee with a "limited attendance potential" for their meetings; an employer who wants to reward a creative and innovative employee, or a library which wants to showcase a new programme or problem solving solution. It is especially successful for the individual who isn't quite ready for a formal presentation at a large conference. The advantages for these persons are enormous. Persons who might never have participated actively at conferences can use poster sessions to take an active role in an informal and non-threatening setting.

First time presenters say that they are impressed that so many librarians are interested in their ideas. From 25 to 200 librarians normally view each of the individual sessions. Feedback is extensive. Many viewers cannot speak to the busy presenters, but correspond with them later.

Presenters may also receive invitations to make a formal presentation at an upcoming conference or meeting, be asked to chair a discussion group based on their session, or be asked to expand on their idea in an article or even a book length treatment of the subject. Presenters say that composing the abstract and putting together the poster session organizes the topic for them so that the majority of the work on such an article or speech has already been done.

Many journal articles are now giving credit to the fact that the original idea first was presented as a poster session. The poster session can also be re-used to accompany a future more formal presentation. The popularity of the format has meant that librarians include poster sessions as part of their promotion for a new programme, an item on their resume, or as a presentation for funding of a government grant or a doctoral thesis. Individual librarians use poster sessions to place their ideas before a large general audience and receive feedback and recognition from the profession for their achievement.

Poster Session Review Process

Part of the process of presenting a poster session is a juried review process. ALA's Review Panel is composed of a chair who coordinates the mailing of abstracts to reviewer and acts as the final arbitrator for acceptance or rejection. Over the past nine years, the Committee, under the direction of Review Panel Chairs, Dr. Ruth Person and Dr. Judith Harwood, has developed an extensive series of "Guidelines for

Poster Session Reviewers."A rating is attached to each abstract. The reviewer is asked to rate the abstract according to a five-point scale based on a list of several criteria. It should fit one of the categories for inclusion:

- recently completed research.
- new and innovative programmes.
- imaginative efforts at resolving a practical problem.

The above subjects may treat any facet of library or information sciences or services. "They should reflect innovation, should be presented systematically, and should incorporate the underlying planning and evaluation process where appropriate," (3) advises the "Guidelines."

Only a handful of poster sessions are rejected each year. There are definite guidelines for those which are rejected. Examples are obvious, such as submissions from a for-profit group such as a software producer with a product for sale under the guise of field-testing research. Exhibit space at conferences is extremely expensive, and crafty individuals are always trying to acquire free space.

Rejected ideas frequently do not meet these criteria:

- Does the abstract explain the poster session clearly?
- What is the concept and how will it lend itself to graphic representation?
- If it is a research paper, is the research completed?
- Is it original work, or has the idea been presented before?

An effort to evaluate should be included in any poster session describing a programme or problem solved.

A poster session of a research study should include the problem under investigation, the methodology used, the major findings and general conclusions. All information should be presented in a coherent and organized manner. The abstract should be accurate. Absolutely no errors should be allowed in the typing of the abstract since the copy must be camera-ready for the abstract booklet.

How are poster sessions presented?

Two guidelines are very important when presenting poster sessions:

- Think in terms of graphic impact.
- Everything should be legible at a distance of four feet.

Much to the disbelief of poster session presenters, there may be more than fifty persons in one booth at any one time. Therefore, legibility is of a prime concern. Primary type or large computer generated fonts should be used. Many poster sessions are now composed completely on computer word processors.

The exceptional poster session has a unified feeling to it, not a pasted-up effect. The display should be well conceived and planned in advance, with attention to visual and graphic impact. Many librarians use the help of professional graphic designers. Lettering and headings should be at least 3/8 inches high. Lettering should be clear and uncluttered. Enlargements enhance a presentation and are especially effective when used to describe a library programme. Effective use of color, particularly in the choice of background materials or fabric can create an attractive poster session.

One problem that the ALA Poster Session Committee has had to address has been the rejection of applications which have required either audiovisual or computer applications as part of their presentations. They have been rejected for two major reasons. The first is that the cost of an electrical hook-up per booth at conferences is extremely expensive, and the second is that computers and audiovisual generated images and sounds can be a distraction and an unfair competition for those without such enhancements. ALA has, in the past through "swap-n'-shops" provided forums for the use of audiovisual and computer assisted displays.

Poster sessions have become a highly effective and well-received method of sharing new and innovative research, programmes, and ideas in the library and information science profession at annual conferences and meetings. They provide an alternative channel and outlet for communication for these associations. They involve individuals who might not have otherwise shared their research with others. They create a Conference-Within-A-Conference where attendees can manage their conference time based on their own perceived information needs. They are effective ways of rewarding innovation and creativity in the profession. Finally, the popularity of poster sessions assures them of a permanent place in the continuing education experiences of librarians of the future.

For more information about poster sessions, contact:

Poster Sessions Committee
American Library Association
50 East Huron Street
Chicago, Illinois 60611, USA

Resources available from the committee include:

- Guidelines for Preparing Poster Session Presentations.
- Poster Session Review Instructions.
- A Proposal to Pilot-Test the Use of Poster/Idea Board Sessions by ALA as an Approach to Conference Streamlining.
- Application Form for Participation in Poster Sessions.
- Star-Studded Ideas for Libraries (ALA Poster Session Abstracts, 1982 to the present time.

In addition, the Committee recommends Jerrold Kemp's *Planning and Producing Audiovisual Materials*, New York, Harper and Row, 4th Edition, 1980. Chapter 16 is especially recommended.

References

1. Stone, Elizabeth W. "A Proposal to Pilot-Test the Use of Poster/Idea Board Sessions by ALA as an Approach to Conference Streamlining, September 1980, unpaged.
2. Comments made to author at annual conference in Dallas, ALA, June, 1989.
3. American Library Association. Poster Session Committee. "Poster Session Review Instructions." unpaged.

EMPLOYER AS TRAINER: THE ROLE OF THE EMPLOYER REGARDING THE IN-SERVICE TRAINING OF LIBRARY STAFF

C.M. Vink
Director, Library Services
Medical University of Southern Africa
70 Ingersol Road, Lynnwood Glen, Pretoria 0081, South Africa

ABSTRACT

Only the employer can train librarians for specific tasks in a specific library. Newly qualified staff and staff without formal library qualifications need special attention and all staff must keep up-to-date with the latest developments. This paper presents the role of the employer in this process.

In-service training is organized training whereby management creates opportunities for workers to become better equipped for present as well as future work. Training by the employer is essential because training by institutions other than the employer can hardly train librarians fully for a specific work situation in a specific library. Newly qualified staff and staff without formal library qualifications need special attention. Furthermore, all staff must be kept aware of the latest developments in the library and information science profession.

The aims of in-service training should be to:

- Provide staff with necessary knowledge.
- Guide them towards greater insight.
- Improve their skills and broaden their experience.
- Develop a positive attitude towards work.

In order to plan an appropriate training programme, the employer must first determine needs. The tasks to be performed in the library must be analyzed and the training needs of the staff taken into account. Both present and future requirements should be taken into consideration.

After the review of the tasks to be performed, the knowledge, insight, skills, and experiences necessary for the work to be accomplished should be recorded. The categories of worker that are to be trained should be noted, and a programme compiled.

In compiling the programme, the goals are formulated. The person responsible for the training is determined, and the group to be trained is selected. Subjects to be covered are decided, the syllabus is drawn up, and the order in which the programme is to be presented is outlined. A decision must be made on the duration of the course and the training methods to be used.

A plan for evaluating the training must be chosen. Training should be evaluated so as to enable one to determine the effectiveness of the in-service sessions.

Who Should Be Responsible for In-Service Training?

Management must accept responsibility for training. The organization under which the library falls, e.g., the university as a whole in the case of the university library, could provide for general training such as training in management and training for supervisors and training officers. The library management should be responsible for training in specialized library subjects and routines and also for general aspects that concern the library.

A committee or an individual should be assigned responsibility for planning and co-ordinating training. The chief librarian or a deputy librarian should also be involved. Libraries with more than 200 employees could consider appointing a full-time training officer. Libraries with less staff should consider making training part of the responsibility of a specific person.

The training officer is not the only person who should be involved in training. A variety of suitable people from within or outside the library can participate as the need arises. All supervisors must take responsibility for on-the-job training. They can also make a valuable contribution to the development of staff by the example they set, the guidance they give, and the standards that they prescribe for staff in the work situation.

Content of training

The employer must determine what the staff have learned at school, college, or university and identify and rectify shortcomings. Duty sheets, job descriptions, should be used to determine what employees must do. It is important to find out what staff do not know about their present or future tasks so that training can be planned to provide the necessary skills.

Some staff members have been trained in library and information science, while others may have no training at all. The training given by different training institutions may also differ. Insufficient education and experience should be supplemented.

Provision is made for general training such as service conditions and administrative procedures. More specialized training is provided for staff who will accomplish these tasks such as cataloging or online searching.

The employer should ensure that workers have necessary knowledge of the environment in which they are going to work. Staff must know the organization and structure of the library as well as its aims and functions. They must know what is expected of them and how to tackle their assigned tasks. Individuals must be aware of the duties attached to their posts as well as the procedures, methods, and policy of their employer. Certain information such as service conditions can only be supplied by the employer because they are specific to the institution.

Training methods

A variety of training methods could be considered. The best method should be chosen for each situation. One method is the use of duty sheets.

A duty sheet is a list of the duties of a specific employee. An employee can be queried about the list of duties. Gaps in the knowledge of workers can be discovered merely by discussing the duty sheet, and this information can be used to determine the contents of the proposed in-service training programme.

An up-to-date procedure manual describes the methods to follow when a specific task must be accomplished. When a variety of manuals have been compiled, these can be used for training. Examples are a general manual including information about the history and functions of the organization, staff information concerning working hours and leave policies. The rules and regulations that apply for the institution can be presented.

A manual for supervisors can help them understand their role and can provide hints for effective management. The reports that supervisors must generate and their other duties should be detailed. Another manual is one that would remain at the issue or charge-out desk to help persons working there to understand circulation procedures.

On-the-job work experiences can improve skills and broaden experience. With this type of training, the employer explains to the employee how a task should be performed, demonstrates how the task is done, and then lets the employee practice the task. For more difficult tasks, ample attention is given until the employees are comfortable with their ability to complete the task accurately and efficiently.

Experience can be simulated in the classroom, but it is only in the practical situation that real experience can be gained. Many people with a library science qualification know very little about many aspects of practical library work. They must therefore be given the chance to gain experience in the work situation under the supervision of an experienced and qualified librarian.

Staff must learn the work they themselves are assigned and must also know about the work expected to be done by their subordinates. One of the disadvantages of inexperienced and recently qualified librarians is that many know little of library routines normally done by unqualified library assistants.

In smaller libraries it should be obvious that every staff member must be able to do the work of others so that the absence of one staff member does not mean that some duties are halted. In larger libraries, it can easily happen that persons working for years in one section can get into a rut. For example, all staff in the cataloging section could work at a service point for a few hours a week to broaden their background. Broader general experience can also lead to a better grasp of situations in a more specialized area. A degree of work rotation is essential if stagnation is to be avoided.

Staff being prepared for a promotion can be prepared using an internship method. This is one method for preparing staff members for managerial positions. It is important to ensure that supervising staff should gain wide experience.

Informal as well as formal training is possible. Even during formal courses, opportunities should be provided for social mixing so that members of the staff become better acquainted, thus motivating staff and helping them develop a positive attitude. Knowledge may be transmitted by means of informal discussions and advice, but use should also be made of written accounts or reports on the work. Subject literature may be read during working hours. Lectures, audiovisual aids, and programmed instruction can also be used.

To promote deeper understanding, staff meetings should be held and visits paid to other libraries or among sections of the staff's own library. Seminars, debates, and discussions of case-studies can also be arranged. Role playing can also be encouraged.

A staff magazine containing social news and work related information is a form of in-service training. If no printed communication is in place, one could be started.

Finally, employers should supplement their in-service training of staff by other suitable continuing education opportunities outside the organization. Symposia and professional meetings are examples of this.

INSTRUCTIONAL TECHNOLOGY IN CONTINUING EDUCATION

Blanche Woolls
Professor and Chair
Department of Library Science
School of Library and Information Science
University of Pittsburgh, Pittsburgh PA 15260, USA

ABSTRACT

This paper defines instructional technology, discusses the use of technology for continuing professional education, and presents suggestions for using CPE programmes to train CPE providers in the use of instructional technology. A brief introduction is given to methods used to develop learning materials.

Instructional technology may have different definitions or different concepts to different people. Explaining the role of instructional technology in continuing professional education requires a common definition for the term, instructional technology. For this paper, the concept is made up of the concepts found in both terms. The first, instruction, focuses on the role of teaching and learning, while the second, technology, becomes the vehicle for implementation. For the world of instructional technologists, instructional technology has been defined as:

> ...the media born of the communications revolution which can be used for instructional purposes alongside the teacher, text-book, and blackboard...It is a systematic way of designing, carrying out, and evaluating the total process of learning and teaching in terms of specific objectives, based on research in human learning and communication, and employing a combination of human and nonhuman resources, to bring about more effective instruction. (1)

All teaching and learning at all levels should be systematic and all methodologies that are applied should be designed based on research in human learning and communication. After any methodology has been applied, it should be evaluated in terms of specific objectives. The unique component of instructional technology in the teaching and learning process is the use of media, combining "human and nonhuman resources to bring about more effective instruction."

This article is divided into three parts. The first explains the use of instructional technology in continuing education. The second part presents suggestions for using CPE programmes to train continuing education providers for the use of instructional technology in planning future CPE programmes. The final part briefly describes methods to develop learning materials using instructional technology. The reader is given citations to additional information immediately following the references.

Using Instructional Technology in Continuing Education

Most continuing education providers are well aware of the use of print and human resources; many are less familiar with the opportunities offered with instruction technology. They may not understand the ease of use and the ways instructional technology can be used to motivate. The use of instructional technology in continuing education requires an introduction to media and its correct use in presentations and programming.

Media is available in a wide variety of formats to entertain, to provide information, and to instruct. These include print, graphic materials, audio and visual media, and programmed instruction. Some media, e.g., print materials, stand alone requiring no additional technology for use. These include books and other graphic materials such as drawings, photographs, diagrams, charts, maps, flipcharts, posters and study prints.

Most other types of media require equipment to project onto a plain surface so that they can be seen by the audience. Among these are overhead transparencies, slides, or filmstrips. Audio media such as recording and audiovisual media such as films and videotapes also require projection and audio equipment for use.

Programmed instruction can be developed in print format when produced in notebook or workbook format, allowing students to learn at their own pace and with a certain amount of choice of alternative sequences called "branching." Programmed instruction provided on microcomputers or through mainframe computers can be very well designed to keep the interest of the learner by branching to more difficult work or by taking the student through remedial work when that is indicated.

Any media format can be used in planning continuing education. Care is taken to match the medium to the message. Once this has been done, the programme can be easily repeated to a second audience. One of the major benefits of providing continuing education using instructional technology applications is the ability to repeat the instruction and assure that future audiences are receiving the same information previously presented. It is up to the continuing education provider to choose the best method for presenting the information. This is the match of media to message.

Each media format offers unique features. Some media show motion, films and videotapes. It is very difficult to demonstrate motion on the printed page. The "seeing" must be translated by the reader. A sequence of sketches may show positions for a swimming stroke, a gymnastics sequence or ballet movement, but a film can show all three in slow motion and then move to a real time performance allowing the viewer to observe over and over if necessary to understand. How to choose the most effective format for the learning experience is crucial.

Books can be read by the instructor, or, with copyright permission, duplicated and distributed. Portions of the text may be transferred to an overhead projector or photographed and turned into slides. Print cannot be heard unless it is read and it cannot show motion. Other graphic materials must be produced in a large enough size to be seen by the audience. If that audience is in a large room or auditorium, it may be preferable to project the image.

When slides and transparencies are used, they must be clearly visible to the audience. At times, the apparent ease of producing transparencies allows presenters to ignore the need for lettering or other symbols in a large enough size on the master to be visible to the entire audience. If the presenter does not want the audience frantically copying the information from the projected image, hand-outs of the contents of visuals should be made available.

While transparencies may be shown in a semi-dark room, slides and films should be shown in a darkened room. Light washes out the image. Television, unless large screen projection is available, can be used only with a very limited audience. The inexpensive duplication of videos makes this a portable format for the audience if they have compatible VCRs at home.

Microcomputers can be used with small groups if the computer is attached to a projector, but this format is most useful with self-instruction or on an individual basis. If the audience is to use the microcomputer, each participant should have access to a computer.

Continuing education providers who are not comfortable selecting instructional technology applications may wish to secure the services of someone who is. As more and more programmes are developed in instructional technology formats, sharing them with others becomes easier and may lessen the need to create again what has been developed and refined.

Novices in the creation of effective training modules must pay particular attention to the quality of any material they choose to supplement their presentations whether it be selected from a commercial source, be created in-house, or be selected from the repertoire of another continuing education provider. It is easy to fall into the trap of pride in production of low quality materials, or succumbing to the pitfall of using an inadequately designed medium because it appears to be all that is available. Certainly providers should not plan to videotape a continuing education programme as it is being given and expect that to be of high enough quality to capture the interest of other audiences. Too often the photographer has merely captured "talking heads."

With all the discussion, it might seem easier merely to lecture, occasionally using a flipchart or a chalk board to illustrate a point. However, once a presenter becomes accustomed to illustrating, appropriately, with media, information becomes more meaningful to the audience. Learners are motivated, and teaching becomes easier for the provider. Again, the ability to use the same media repeatedly can assure that subsequent audiences are given the same information as given to previous audiences.

For those persons who are not skillful users of instructional technology for continuing education, one of the goals for personal continuing education might be to learn how to use instructional technology appropriately and efficiently. Training continuing education providers can be accomplished through continuing education experiences.

Training Continuing Education Providers to Use Instructional Technology

Instructional technology professionals are taught how to develop training modules. They learn to create, develop, and use instructional technology in their master's and doctoral studies. They improve their skills through on-the-job practice. To replicate this in-depth training in brief, continuing education experiences is impossible. However, continuing education providers learn to use media and use it very effectively. Choosing the media for the message is a first step. Primary application of each type has been presented earlier.

Continuing education providers using instructional technology in presentations must be aware that the environment will accommodate the medium. As stated previously, a room that cannot be darkened will make the most exciting film or slide ever produced into an ineffective experience. Any item used must be clear and visible to the entire audience.

Electricity is needed to run most equipment. One electrical outlet 100 feet from the speaker may halt the use of more than one format of media because switching from one piece of machinery to another would be awkward. If no extension cord is available, it becomes impossible. Other problems include bulbs that do burn out, equipment may not be compatible for the software, or a slide carousel may not be available. For reasons such as these, many audiovisual experts believe in the principle that "Anything that can go wrong, will go wrong". Nevertheless, media and equipment carefully chosen and presentations carefully planned will increase the success and improve the audience's understanding.

Often commercial materials are not available for the learning experience. In this case, it may be preferable to create materials that would be directly applicable to the training.

Developing Continuing Education Learning Materials

The introduction to *Techniques for Producing Visual Instructional Media* states:

> ...the purpose of any worthwhile instructional material should be to communicate quickly, clearly, and memorably: thus, materials should be prepared and used only when they will do the job of communicating better and quicker than any other method of communication available. (2)

When materials are prepared, there are guidelines that should be followed. Guidelines were developed for the Continuing Library Education Network and Exchange (CLENE) and were published as Concept Paper 1 (3). Two sections, "Planning Continuing Education Materials for Local Use" and "Planning Continuing Education Materials for Wide Dissemination," are reproduced in part and with revisions in the section that follows.

For Local Use

In planning for the preparation of continuing education materials for local use, many factors must be considered: audience, training needs and methods, teaching and learning activities, and evaluation. The following six items deal with the details which have been selected as top priorities for the preparation of continuing education materials.

The specific audience should be determined

Rationale: Differences in training needs, methods, and vocabulary will occur in the proposed training programme depending upon the audience. *Example*: Is the audience made up of experienced librarians? beginning librarians? paraprofessionals? library patrons, senior citizens or school students? trustees?

Define training needs

Rationale: The training needs of the audience must be determined. Some assessment must be done to determine if the audience NEEDS training. *Example:* If the potential audience is one person and the training requires a simple sentence of instruction with opportunity for question/answer, the instruction is unlikely to be repeated to a wide audience. In this instance it would be ineffective to develop a sophisticated training resource for this simple need.

If training will recur frequently to an ever changing audience, and consistency of training is desired, a decision to produce a training package may be appropriate.

Search for available materials

Rationale: Suitable materials for specific training needs may be currently available. If materials are available, unnecessary duplication may be costly.

Example: Another instructional product may be on the market which could be adapted to the training objectives.

Seven general sources of information about instructional products include:

1. Bibliographic information sources.
2. Periodicals which review instructional products.
3. National, regional, state or local agencies which develop, coordinate, or disseminate instructional materials
4. Training programmes which make use of instructional materials and develop bibliographies as part of published outlines.

5. Catalogues of publishers.
6. Libraries which have extensive collections described in published bibliographies.
7. Personal contacts (those persons successfully using instructional materials may recommend effective materials).

Determine learning methods and format of programme

Rationale: A variety of experiences will lead to learning. Among these are discussion, reading, writing, listening, observing or viewing, demonstrating, imaging, and evaluating. A large number of media possibilities exist to provide these experiences: books, magazines, programmed materials, pictures, transparencies, microforms, chalkboard materials, displays, recordings, filmstrips, films, videotapes, microcomputers, and television. No one format is best for all purposes. The choice of format must be consistent with training objectives and media capability. *Example:* The production of a 16 mm film to train cataloguers with new Library of Congress rules would be an expensive and unsatisfactory way to conduct training in comparison to less expensively produced and easy to edit slides or transparencies. The film is inappropriate for the content of the material. On the other hand, a short film or videotape may be an excellent means of acquainting the library patron with the floor plan of the library.

Choose method of production and plan for production

Rationale: This is the *raison d'être* for this section. If a potential producer is not an expert in the technology of actual production of the various formats of training materials, such skills can be learned in production and instructional design courses at colleges, community colleges, and high schools. Textbooks or manuals exist explaining the techniques of graphic design. Such manuals are listed following the footnotes. Or, the producer may choose to work with an experienced technician.

Six steps are outlined below for the production of instructionally effective materials. They include:

1. State the purposes to be served by the chosen topic.
2. Identify the target audience and describe the characteristics. Lists any required entry level behaviors or competencies prerequisite to using the materials effectively. The producer might wish to develop a pretest to determine the present level of competence of the potential user with the subject. This could become a part of the learning packet.
3. Identify the specific instructional objectives which the material is to accomplish.
4. Relate the above objectives to the subject content and purposes of the material being designed. Indicate how the content and purposes will lead to the objectives.
5. Describe the procedures, teaching and learning activities, treatment, and instructional resources necessary to accomplish the objectives as stated.
6. Provide a means to evaluate learner performance in terms of the accomplishment of the objectives.

Determine if materials produced could be used with ALTERNATIVE audience.

Rationale: Materials produced for one audience might be easily adapted for another audience. *Example:* The head of technical services may perceive the need to acquaint members of the department with new Library of Congress cataloguing rules. The target audience in this case are professional cataloguers and the general objective is to help them develop the ability to catalog using the new rules. Materials developed (overhead transparencies or slide/tape presentations) might be appropriate for a beginning cataloguing class (perhaps with a different explanation on the accompanying audio tape) or for persons working in clerical positions in the cataloging department (with still another audio tape). In this way materials developed for continuing education audiences could be made instructionally effective for different audiences of learners.

For a Wide Audience

When planning for the production of continuing library education materials which could have wide dissemination, the following must be done:

1. *All* materials protected by copyright or other types of restrictions must be released from the owner; and
2. *All* persons who appear in the production may need to sign talent releases. In the United States, persons who appear in such audiences must sign a release form.

During the production of instructional materials, nonprofessional "actors" or "actresses," such as library patrons, students, teachers, and other persons may be involved in the production as "talent." *Anyone* who appears in a photographic image or is heard on a sound track of any product, whether still, motion picture, audio tape, videotape, or television, must sign a formal release for the use of that photograph or audio recording for instructional purposes. This applies to images or sounds which will be reproduced for others, or for use in the original form in other than a very narrowly prescribed situation. The checklist below may help determine if the planning process has been completed.

- Assess the need for the material?
- Purposes to be served by the chosen topic include:
- Identify the target?
- Identify the specific instructional objectives?
- Related the above objectives to content and purposes of the material being designed?
- Identify entry level competence?
- Described the procedures, teaching and learning activities, etc.?
- Determined the procedures for evaluating the learner's performance in terms of the objectives?
- Obtained releases from the individuals who are to appear as "talent" in the production?

The use of media in its many formats can serve as a unique educational resource for continuing education providers. With careful planning, the ease of using media and the capability of duplicating the training session for other audiences make this a necessity. Practice with using media can assure the provider that this is true.

References

1. United States Commission on Instructional Technology. *To Improve Learning: A Report to the President and the Congress of the United States*. Washington, D.C.: U.S. Government Printing Office, March, 1970.
2. Introduction to Minor, Ed and Harvey R. Frye, *Techniques for Producing Visual Instructional Media*. New York: McGraw-Hill Book Co., 1970.
3. Woolls, Blanche and Brooke Sheldon, *Developing Continuing Education Learning Materials*. Washington, D.C.: CLENE, n.d.

Bibliography

Mether, Calvin E., *et al. Audiovisual Fundamentals: Basic Equipment Operation, Simple Materials Production.* 4th ed. Dubuque, Iowa: Wm. C. Brown Publishers, 1989.

Raines, Claire. *Visual Aids in Business.* Los Altos, California: Crisp Publications, Inc., 1989.

Satterwaite, Les. *Graphics: Skills, Media, and Materials.* 3rd ed. Dubuque, Iowa: Kendall/Hunt Publishing Company, 1982.

Volker, Roger and Michael Simonson. *Media for Teachers: An Introductory Course in Media for Students in Teacher Education.* 4th ed. Dubuque, Iowa: Kendall/Hunt Publishers, 1984.

MARKETING CONTINUING PROFESSIONAL EDUCATION

Darlene E. Weingand
Associate Professor and Director
Continuing Education Services
University of Wisconsin-Madison,
School of Library and Information Science,
Helen C. White Hall, Room 4623, 600 North Park Street,
Madison WI 53706, USA

ABSTRACT

This paper expands seven major points in the definition of "marketing." These points highlight the managerial quality and usefulness of the marketing process in relation to CPE.

Introduction

Marketing is a process of exchange. It is a relationship that provides mutual benefit to both parties in the transaction. As an evolving process, it is influenced by definitions, perceptions, environmental and cultural conditions and trends. In the past, the concept of "selling" has often been equated with marketing; this has now changed fundamentally. Today's interpretation of marketing is a full 180 degrees from the "selling" definition. Rather than the design of a product which is then "sold" to a potential public (whether they want it or not!), marketing now involves an approach to product design that reflects the documented needs of the target population.(1)

The traditional view of librarians toward marketing has been in terms of promotion only, and promotion in the sense that the libraries' products are so intrinsically valuable that people "should" use them and therefore only need to be alerted to their existence. In addition, since promotion has tended to focus upon programmatic activities, such as film showings and story hours, rather than upon the full range of services, this incomplete approach has not succeeded in informing the community at even a partial level. Most clients have a very poor sense of what they have a right to expect, and demand, from libraries.

Such a limited scope and perspective has cheated both libraries and communities over time, as libraries are tremendous community resources which have been dramatically underused and consequently undervalued. Operating in the realm of "intrinsic value" as opposed to the identification and satisfying of client needs, libraries have struggled valiantly to survive in years of economic hardship and, today, are beginning to lose the struggle. The economic health of libraries, if not survival itself, may well depend upon the use of marketing and planning strategies to alter significantly the perceived role and placement of the library in society.

Continuing professional education suffers from similar attitudes and perceptions. Like parenthood and one's national flag, continuing education is viewed as worthwhile and important in general terms. While continuing education is recognized as having intrinsic value, often in terms of others rather than self, making the connection between a professional's individual needs and the appropriate educational activity is all too frequently a scatter-shot endeavor. In many cases, the necessary prod is the need to become re-certified, the unknowns presented in a new job or set of responsibilities, or administrative fiat.

Events reach a point of crisis, however, when the concepts of occupational obsolescence and coping with a changing profession are brought home to the workplace, fostering an atmosphere of anxiety and concern. Laissez-faire attitudes toward continuing education now have the potential to become more personalized; it is at this point that an awareness of mutual benefit can be nurtured and encouraged to flower. It is the "teachable moment" when need can be expressed, but is there anyone to hear?

Marketing: The Ingredients

Kotler offers the following definitions of marketing:

> Marketing is the analysis, planning, implementation, and control of carefully formulated programmes designed to bring about voluntary exchanges of values with target markets for the purpose of achieving organizational objectives. It relies heavily on designing the organization's offering in terms of the target market's needs and desires, and on using effective pricing, communication, and distribution to inform, motivate, and service the markets.(2)

There are seven major points to be emphasized in this definition:

1. Marketing is a *managerial* process involving analysis, planning, implementation, and control.
2. Marketing is concerned with *carefully formulated programmes*, not random actions, designed to achieve desired responses.
3. Marketing seeks to bring about *voluntary* exchanges.
4. Marketing selects *target markets* and does not seek to be all things to all people.
5. Marketing is directly correlated to the achievement of *organizational objectives.*
6. Marketing places emphasis on the target market's (consumer's) *needs and desires* rather than on the producer's preferences.
7. Marketing utilizes what has been termed the "marketing mix" or the "4 Ps": *product, pricing, place/distribution*, and *promotion/communication.*(3)

These seven points highlight the managerial quality and usefulness of the marketing process. A closer examination in terms of continuing professional education will serve to clarify the relationships.

Marketing is a MANAGERIAL process involving analysis, planning, implementation, and control. Effective continuing education does not just "happen." Even if the activity has a designated staff, a committee or other group involved in development and production, there needs to be an individual with overall managerial responsibility who can guide the processes of analysis, planning, implementation, and control. These four processes are critical to successful programme development.

The first, analysis, focuses on the internal and external environments. Commonly termed a "marketing audit," this analysis phase critically examines limitations, opportunities, strengths, and trends within the programming group or organization and beyond, into the micro and macro external environments. Both secondary sources of already collected information and primary source data gleaned from interviews, telephone or mail questionnaires (or related methodology) are used in this close examination. The resultant data provides the necessary resource information upon which to base informed decisions.

Marketing is concerned with CAREFULLY FORMULATED PROGRAMMES, not random actions, designed to achieve desired responses. As a result of the information gained in the marketing audit, programmes are designed that will address the identified needs of the target populations. Some of these needs will be perceived needs or desires, i.e., needs which are recognized and expressed by the target groups. Other needs may be real needs, identified in the audit and a concern of the planning committee, but which have not as yet reached the consciousness of the desired audience. This is the more difficult arena in which to operate, as worthy programmes may be planned and presented which do not draw the desired numbers of attendees. In this unhappy situation, it is frequently necessary to table the programme and wait until an awareness of need becomes evident, or, in some situations, an educational effort directed toward producing this awareness may be mounted. In any event, whatever programmes are formulated must be correlated to measurable objectives so that evaluation procedures will be meaningful.

Marketing seeks to bring about VOLUNTARY EXCHANGES. There is considerable controversy concerning the merits of mandatory continuing education. Certification and re-certification requirements occur frequently in elementary and secondary school situations, in some governmental jurisdictions regarding public library personnel, for medical librarians and, just recently, for archivists. But whether this type of structure is in place or not, the actual attendance for a particular programme or course may still be voluntary (unless, for example, a supervisor tells an employee, "You don't manage your time well; attend this course on time management.")

In the vast majority of situations, however, attendance at a specific programme is voluntary and, as such, requires that there be a sense of partnership and mutual benefit. The programme developers seek to provide an effective continuing education opportunity that is well-attended; the participants seek to engage in the selected learning experience in order to achieve personal and professional goals. The ideal scenario is that both sides are satisfied.

Marketing selects TARGET MARKETS and does not seek to be all things to all people. No provider of continuing professional education can hope realistically to attract everyone, nor can any given programme be meaningful to the widest spectrum of potential attendees. Therefore, as part of the planning process, specific target markets need to be identified and each programmatic event designed with specific target groups in mind. In other words, a course directed to managerial personnel will be very different from one that focuses on support staff. A course which helps select an automated system will attract a very different audience than a course featuring children's literature. Promotional materials, too, should be targeted to specific markets. This specificity has the dual effect of keeping costs down and more effectively reaching the desired clientele groups.

Marketing is directly correlated to the achievement of ORGANIZATIONAL OBJECTIVES. Both the objectives of the providers and the objectives of the specific programme must be considered in the programme design. As stated above, measurable objectives for the programme itself facilitate the gathering of meaningful evaluation data. But beyond the individual programme lie the objectives of the provider group.

The effective marketer recognizes that a unique market share must be carved out of the total possibility range. The continuing education provider must ask, "What can I do that is better, faster, cheaper, or otherwise more effective than my competition?" This, of course, is predicated on the assumption that the competition is known, and this should be the case if a thorough marketing audit has been done. When the provider creates continuing education opportunities that are within the scope of the identified market share, the results are far more positive for both producer and consumer.

The provider's mission is the philosophical base which supports the entire marketing effort. Written in sand and not stone, the mission is a simple statement of purpose and intent that is designed to communicate what the continuing education provider does and why to both the internal staff and the external community. The mission is the first swamp that many managers stumble into, and is more often than not comprised of simple platitudes that say little and mean less. All too often, staffs are not clear as to what the provider's role is vis-a-vis the community.

Conversely, the provider that is proactive, seeking out and responding to client needs in an accurate and timely fashion, is also a provider that probably communicates in a two-way mode with its target markets, i.e., doing a good job of establishing high quality public relations.

Marketing places emphasis on the target market's (client's) NEEDS AND DESIRES rather than on the producer's preferences. All too often, continuing education providers seek to produce programmes in a format, time frame, or other parameter that relates more closely to producer preference than client convenience. Provider strengths and limitations are legitimate considerations, but the focus must always be on the client. What topics, delivery system, time of day, day of week, fee range etc., most closely approximate the identified client needs?

Marketing utilizes what has been termed the "marketing mix" of the "4 Ps": PRODUCT, PRICING, PLACE/DISTRIBUTION, and PROMOTION/COMMUNICATION. Kotler formulated this mnemonic device to aid in memory retention of these important concepts. The "4 Ps" are as applicable to continuing professional education as they are to any product, whether for profit or nonprofit. Considering each of these elements of the marketing mix separately in an educational context, the following discussion ensues.

The *product* in continuing professional education is the educational experience. Once educational needs of designated target markets have been identified, the design of the product can commence. Design is, of course, closely correlated to educational objectives for the experience with a programmatic eye always focused on desired outcomes. The "Guidelines for Quality in Continuing Education for Information, Library and Media Personnel," published by the Continuing Education Subcommittee of the Standing Committee on Library Education (SCOLE) of the American Library Association in December 1987 provide clearly stated criteria for quality in the development of continuing educational activities.(4)

The concept of *pricing* in continuing education has two foci. First, the knowledge of what a continuing education activity will actually cost the provider to produce, and second, the development of a participation fee that will adequately cover both direct and indirect costs. Pricing is like walking a tightrope; sufficient knowledge must be present to determine a fair fee, yet that fee cannot be either too great or too small. Too large a fee will discourage registration; too small a charge will cast doubt upon programme quality. There is also the risk that a programme will not recover the necessary costs. In general terms, moreover, there is no profit in the provision of continuing education. It is a service operation and, as such, seeks to reinvest any monetary overage in future programme development.

Place/Distribution: the delivery of continuing professional education, is becoming more varied with each new technological development. No longer need instructor and student be in the same physical location, or even in the same time frame. Of course, in-classroom situations still exist, but opportunities are arising in a multiplicity of formats and delivery systems. Today, continuing education can be found in courses transmitted via satellite, broadcast television and radio, teleconferencing networks, correspondence mechanisms, and the list continues to grow. Technology has had the result of discounting time and space, allowing study opportunities that more closely correspond to the student's daily responsibilities and convenience. The advent of the videocassette recorder has opened the door to time-delayed interaction with television.

Some courses are transmitted via the vertical blanking interval, those unused data lines in normal TV broadcasting, interactive programmes combine the special attributes of optical technology and computerization. There appears to be no end to creative uses of technology for educational purposes. However, as with product design, the decisions regarding choice of delivery system must also be grounded in the educational and convenience needs of the target markets. Further, issues of quality are also present in the selection of modes of distribution and require careful consideration.

Promotion/Communication is the last "P." Once the target markets and their needs have been identified, the products and channels of distribution have been designed, and an equitable fee for all concerned has been assigned, then, and only then, promotion of the continuing education opportunity begins. Promotion must be viewed as communication: the sending of a message from continuing education producer to potential students concerning what is available. Since messages are sent from sender to receiver through a channel, the medium of transmission should be selected with the focus of attention still on the target market. Questions can be asked, such as, "Where does my desired audience go for information? What media gain their attention? Where can I promote that my audience is likely to frequent?" In other words, the strategies of promotion are designed with the client continually in mind.

Marketing can also be viewed as the effort to enhance the image of the organization in the eyes of consumers or other publics. A provider needs to position an overall image in the local community and regional market in much the same way as does any other business. In reality, "public relations" can be an accurate descriptor for an effective marketing operation: the relations that an organization has with its various publics or target markets.

Marketing: Overcoming Potential Problems

Using marketing as a managerial process as outlined above can enhance the positive aspects of continuing professional education and forestall the potential difficulties. And there can be difficulties. Stone cites a number of common marketing problems facing providers of continuing education. These include:

- Too few users.
- Declining market.
- Declining utilization in specific areas.
- Changing clientele.
- Poor image.
- User dissatisfaction.
- Insufficient funding.(5)

Stone further asserts that providers respond to these problems in three different ways, essentially saying:

- "We have a good product, people will want it, and support it."
- "There are a lot of people out there who might want our programme ...We must .go after them."
- "We must develop our continuing education activities and products to serve the needs of our target markets."(6)

The first of these statements portrays the traditional view of librarians toward the intrinsic worth theory of libraries, the "oughts" and "shoulds" referred to at the opening of this chapter. The second statement illustrates the "selling" approach to marketing which, while once in vogue, is now yesterday's methodology.

It is the third statement that reflects the tone and substance of today's marketing principles and strategies and truly represents the appropriate relationship between the continuing education provider and the target audience. It is marketing that serves as nurturing soil for product development, the fertilizer and gentle rain for product growth, and the arrangement of final blooms as represented in a well-conceived and implemented continuing educational opportunity. Potential problems can be averted, potential successes can be enhanced by injecting the process of programme design with appropriate spurts of marketing power.

Summary: The Process of Marketing

In examining again the concepts described above, marketing can be summarized as a process that includes six major components:

- *The marketing audit*...a snapshot in time of external and internal environments with an eye toward identifying strengths, limitations, threats and opportunities, an assessment of needs in both the provider and the target markets served; the collection of data in order to determine a unique market share.
- *The product*...the educational opportunity, the range of which is directly related to organizational mission, goals and objectives that are grounded in the needs and convenience of the target market.
- *The price*...thecost required to produce a product, including both direct and indirect costs, the fees that are charged to cover costs.
- *The place*...how the product will be distributed; the channel that links product and consumer.
- *The promotion*...communication of information and benefits to target groups.
- *The evaluation*...the monitoring of how the marketing/planning process is progressing and a final assessment of how well everything worked.

Marketing is important; it is a managerial tool whose power has been largely unrecognized in the nonprofit sector while being greatly exploited by the profit sector. While it is fair to state that a naive approach to marketing is the rule, not the exception, it is time to become savvy in the ways and nuances of marketing practice in order to become competitive in an increasingly competitive world.

The world is a rapidly changing place. Continuing professional education providers that seek not only to survive, but to survive in a successful and meaningful way, must continually monitor the pulse of that change, along with current and potential target markets to be served. It is a challenging time, a time in which library and information professionals need continuing education to an increasing degree. Providers have an unparalleled opportunity to become both instrumental and essential to the changing society. Marketing can be the catalyst to make things happen; it is the key to a dynamic future.

References

1. Weingand, Darlene E. *Marketing/Planning Library and Information Services*. Littleton, CO, Libraries Unlimited, Inc. 1987, 3.
2. Kotler, P. *Marketing for Nonprofit Organizations*. Englewood Cliffs, N.J.: Prentice-Hall. 1975, 5.
3. Weingand, Darlene E. *The Organic Public Library*. Littleton, CO: Libraries Unlimited, 1984, 39.
4. The "Guidelines..." are available from the American Library Association, 50 East Huron Street, Chicago, Illinois 60611.
5. Stone, Elizabeth W. Marketing Continuing Education. In: *Continuing Education for the Library Information Professions*. William G. Asp *et al.*, Hamden, CT, The Shoestring Press. 1985, pp. 108-109.
6. Stone, Elizabeth W. 109.

MARKETING TECHNIQUES APPLIED TO THE DEVELOPMENT AND PRESENTATION OF A SEMINAR PROGRAMME

Margaret Gallacher
Managing Director and Principal Consultant
Information Consultancy Services
P.O. Box 17184, Kardri, Wellington, New Zealand

ABSTRACT

This paper describes the marketing techniques to attract participants to a yearly continuing education and staff training programme for librarians and managers of information services. The planning strategy and development of the workshop are described.

Each year, a continuing education and staff training programme is developed for librarians and managers of information services. The objective has been to assist librarians to maintain their professional expertise and competence in managing and developing library and information services and to provide them with learning opportunities for career enhancement and job enrichment.

The programme includes seminars and workshops about financial planning, budgeting techniques, services marketing, communication and planning for automation. An average number of 15-20 library managers have attended each workshop comprising skills training activities and group discussions. The benefits to the participants have been introducing them to other disciplines, "hands-on" experience, and learning by exchanging ideas with professional peers.

The programme was first developed after an extensive market research exercise. This was designed to assess the demand for the programme from newly qualified and experienced librarians working in public, special, and academic libraries. These librarians were queried about the subject areas of most interest to them and their expectations of the standard, style, cost and features (duration, presentation methods, caliber of presenters) of the programmes.

The conclusions reached from this initial survey were that, although most of the sample affirmed the need for a continuing education programme, it more likely would be supported by experienced librarians filling management positions. In particular, they were seeking opportunities to develop their management skills in the areas relating to business planning, budgeting, financial management, marketing, and communication.

In order to market the programme, an orthodox action plan was prepared and set in place. This included identifying the size and segments of the library market and assessing the economic influences on the level of demand for continuing education services and the effects of other competition groups offering seminars and workshops to the market.

It thus was decided to hold an annual programme of two to four seminars of either one or two days in length, with each to be led by presenters with expertise and practical experience in specific subject areas. Senior and experienced librarians and library managers from special and public libraries were to be the target groups for participants. The fees for attending were to be calculated to recover all costs and provide a small profit margin. Each seminar programme was to be designed in such a way that it could be held in any New Zealand city where there was a range of well-established libraries and information services.

The size of the market was assessed by calculating the type and number of libraries in regional districts throughout New Zealand. Also calculated was the number of librarians holding managerial and senior positions.

One important dimension to the marketing of the seminars was the economic reforms undertaken in New Zealand since 1984. These have had considerable effects on many organizations and created a widespread concern that all business operations, including information systems and services, are managed very efficiently to ensure that they remain cost effective. In addition, there have been greater expectations of librarians as information specialists increasingly responsible for developing policies for information services, for managing financial resources and marketing the services so they support the business goals of the organization. The seminars, therefore, were designed to emphasize the management functions of librarians.

Part of the overall planning strategy for the programme included assessing the degree of competition from other staff training agencies, management consultants, and technical colleges and university schools of librarianship. It was decided that, since none of these groups were offering continuing education seminars of special interest to librarians and information managers, there was the potential to develop a market for this product. The seminars, therefore, were to be well organized, relevant and practical in their orientation and structured to provide librarians with the opportunity to share and exchange ideas and debate current issues.

Thus, the influences on the decision to market a continuing education programme related to determining the rate at which librarians would attend seminars, assessing the impact of competition from other training and management organizations and countering this by rigorously defining the skills and educational needs of librarians in order to develop a unique product. In addition, consideration was given to New Zealand's current programme of economic reform and organizational changes and the influence these factors were having on librarians to develop their managerial skills and expertise.

The marketing mix applied to the development of the programme included defining the characteristics of the product, range of subjects, recommended size of group for each workshop, duration, format and style of presentation, and anticipated benefits to participants. The pricing and communication and promotion mix to be adopted was determined. All seminars were to be evaluated in terms of the response rate from the market and the effectiveness of the programme content itself as perceived by the participants.

The principal goals of the communication mix were, and currently are designed, to highlight the seminar's benefits in an informed way. The programme content is differentiated from that of competitors. A sound reputation is established for presenting quality and practical seminars.

Once the market research is completed for each year and the content of the seminar programme is defined, the calendar of dates for each workshop is established. Personal selling, advertising, and indirect forms of promotion have been considered appropriate means of marketing the programme. A brochure describing the details of the seminar calendar is mailed directly to library managers requesting them to register their interest. Advertisements also are placed in professional journals.

Four weeks prior to each workshop, a brochure describing its content and incorporating an enrollment form is mailed directly to library managers and staff training personnel of large organizations and institutions. The objective is that advertising is to be supported by the timely distribution of sales literature and enrollment forms so that librarians are encouraged to plan for their own involvement in the continuing education programme.

One very important requirement is that mailing lists are kept up-to-date and in addition, where possible, contain accurate details of the names and designations of all librarians.

Within two weeks of each workshop, follow-up telephone calls describing in more detail the programme content and its application are made to ensure that enrollments are received in good time.

An evaluation is conducted each year. Participants at each workshop are requested to make a written evaluation of the quality and relevance of its content and the standard of presentation. Additionally they rate the overall success on a 1-5 points scale.

The seminar programme has been successful in attracting a pleasing number of participants from a large proportion of New Zealand special libraries. Over the last two years, eight seminars have been held in the main cities with more than 120 librarians from 75 libraries participating. More special libraries have participated in the programme; public librarians have not shown the same interest nor evinced the same degree of commitment to continuing education programmes.

All seminars have been highly rated because of the practical nature of their content and its direct application to library management. The caliber of the workshop leaders and the enjoyable atmosphere generated by a well-organized and well-managed programme have been noted. Additionally, participants have commented on the comfortable surroundings for the seminars.

The policy for presenting the programme has as its prime concern to identify rigorously the specific needs of librarians and library managers by directly involving them in the development of the seminars. The marketing strategy for this product emphasizes extensive direct contact with key client groups and includes stringent attempts to monitor and evaluation participation rates by librarians and their responses to all aspects of the seminar programme itself.

ADULT LEARNING IN STUDY CIRCLES (1)

Henry Blid
Brunnsvik's Folk High School
P.O. Box 740, Ludvika S-771 01, Sweden

ABSTRACT

"Adult Learning in Study Circles" takes a brief look at the historical development of "folkbildning" and study circles. It emphasizes the three beneficiaries of the circles, that is the individual participants, their organization, and the government. It discusses the approach used by the circles: their democratic principles, the use of a conversational mode instead of discussions, the importance of books for the learning, the need for planning toward action and change, and, finally, the role of the study circle leader. The paper makes reference to study circles in Sweden and in the United States.

Study circles: this phenomenon forms an integral part of the culture of Sweden and the everyday life of its people. But the circle approach is common in the other Nordic countries and used all over the world.

The conceptual idea developed outside Sweden. England and its Oxford Movement were often mentioned as being the source. Nowadays we are close to a consensus that the study circle has American roots and that the ideas can be traced back to the Chautauqua Literary and Scientific Circles (2) which reached huge numbers in the United States in the latter part of the 19th century.

During that period Sweden experienced the transfer from a backward, quite undeveloped country toward a modern society. But unlike most other countries in Europe, Sweden did not proceed from feudal or semifeudal conditions but from a mainly agricultural economy, to a large extent based on owner-farmer and freeholders. Historically, these farmers and freeholders have always been powerful. They were part of the old Swedish parliament as were the clergy, the nobility and the bourgeoisie. The Swedish parliament is, in fact, the only parliament in Europe where owner-farmers have served as members since the 16th century. As a result, the positions of the other classes were comparatively weaker.

The rapid population increase during the 19th century and its accompanying influx of village people to the cities resulted in the cities having many inhabitants who were descendants from farmers and peasants. It should be noted that the illiteracy level in Sweden may have been comparatively low at that time due to, among other things, the 1842 legislation on primary schools.

Undoubtedly, the new urban inhabitants carried village traditions and attitudes to their new environment, of which the implications were observed first quite recently. To these belong strong feelings of joint responsibilities and social unity (3). The increase of popular movements, nonconformist churches, temperance organizations, labor movement organizations and cooperative societies cannot be fully understood unless a village tradition of solidarity and loyalty is taken into account.

These new popular movements struggled to maintain and develop their organizations. Their need for skilled board members for positions such as chairmen, treasurers and secretaries is a certified fact, as is also the reluctance of the "educated masses" to assist. The "uneducated" maintain a reciprocal suspicion. This mutual dislike can still be seen between workers and university graduates in Sweden. It prevented the establishment of links between the Swedish universities and the popular movements, unlike the development in England and the United States.

The Swedish idea of "folkbildning" (perhaps best defined as the growth and development of individuals *and* organizations free from any school system) developed in this environment and gradually led to the establishment of the modern study circle movement in 1902. The concept of learning in study circles cannot be discussed unless this background is taken into account. The reliance by most study circles on books and self-directed learning rather than on the teaching by any expertise, such as teachers, can, of course, be interpreted as an echo of the Platonic belief that people can learn, reason and act intelligently without being told what to learn, what to think and what to do. But the development seems to be more of a refusal by ordinary people to succumb to any paternalistic intentions, however benevolent these may be, and of an ambition to stand on one's own feet. Actually, this attitude reflects traditional Swedish views on human dignity. The relatively strong positions of the organizations of the handicapped and the meagre influence exerted by the Swedish charitable organizations give additional evidence of this.

From the very beginning, learning in study circles has had two beneficiaries: the individual learners and their organization. They depend upon each other. The individual participants emerge from the learning experience more resourceful than before (4). This means that the members not only become better informed but also more active and, consequently, make their organization more active and forceful. The American International Union of Bricklayers and Allied Craftsmen considers the study circles to be a key to a stronger union:

> The study circles are a means of expanding the education of the members, helping them to understand the Union, the industry, and groups and institutions outside our Union important to our future... We are looking for members' best judgments which can come from open discussion with fellow members. We want all members to speak up, to be challenged, to respect the opinion of others, and to grow from experience. The result can be more informed, active, and self-confident union members and citizens, with renewed pride in their crafts, their union, and their stake in their organization's future (5).

The study circles decide on their business but are simultaneously conscious of a loyalty to their organization. They provide the organizations with well-informed members and important sources for future leaders. This results in vital organizations, well prepared for the future and capable of taking good care of the interests of the members.

Although stiff formal patterns are avoided, democratic principles should always govern the study circle meetings. These principles go beyond the formal requirement "one person, one vote" by demanding that the opportunities to participate in the conversation should be equally distributed, and that those who are shy or have difficulties in participating should be encouraged and helped. Democracy cannot be learned from textbooks but by actually practicing it together with other people. Neither can human understanding be achieved solely through reading. Social and democratic learning are accomplished as an integral and necessary part of the social interplay of the study circles. Furthermore, the climate for learning should be inquiring and inquisitive, not contentious; respectful, not pressuring.

Any learning in any subject or topic following these principles cannot but strengthen the democracy of the organization. This strengthening has further implications. it is, for example, agreed that the work of the Swedish study circles during the 1930s and 1940s actually prevented the totalitarian forces from achieving any influence worth mentioning. Realizing this, the government increased its funding of the study circle and requested the organizers of the circles, the national educational associations, to extend their activities from member organizations to the public. Thus, we can say that the Swedish society became a third beneficiary of the study circle approach.

Study circle learning occurs in groups of people who can freely exchange experiences and knowledge with each other. People voluntarily come to the study circles. They bring with them varied personalities, experiences and backgrounds. They give different ideas, and are encouraged, often requested, to look at issues from different viewpoints. In this way, they learn from each other. The single-mindedness of most textbooks and the one-sided subjectivity so common among teachers are counteracted and balanced, as members are exposed to the different views and experiences of other participants in the study circle.

Learning in study circles consists of integrating facts, values, ideas and experiences, thus creating a cobweb of links and connections. The values and attitudes of the individual members are sometimes only passing whims. Sometimes they are so deeply rooted as to appear to be innate. On a closer examination, these "innate" attitudes most likely consist of subtle cultural inheritances resembling Jung's archetypes. The circle learning process links these elements not only to each other but also relates the individuals to his fellow-learners and their environment, the present as well as the past ones.

A former director of studies at the Swedish Workers' Educational Association (ASF) once described the learning process in study circles in a small pamphlet about the art of conversation. He identified three necessary steps:

- Preparations at home by all individual participants *before* the session.
- Self-activity by all *during* the session.
- Self-activity *after* the session alone at home with books and notes.

He discarded discussion as a mode of interaction. Instead he preferred conversation, an exchange of views and information in a relaxed manner. "Is it possible for two or more ignorant people to talk so they can learn something out of it?" he asked, giving his answer as follows:

> The conversation should be a means to make every participant spiritually active. To talk is the same as to "think aloud." To participate in a conversation is to think aloud under control. One has to make one's knowledge clear by expressing it in the conversation. One has to revise and correct what is erroneous by making comparisons with other people's experiences (6).

Learning in study circles includes, however, more than the individual acquisition of new knowledge, new ideas, new knowing, skills and attitudes and of unlearning, that is, the active discarding and forgetting of what is obsolete. It is a creative, human development process aimed at changing individuals and their environment. Moreover, it goes beyond the notion of individual, self-directed learning since people in study circles are collectively involved in an intensive interactive learning.

This collective process brings people closer to each other, improves their ability to cooperate and develops their feeling of responsibility towards the group which they have formed with their fellow-learners. The small number of participants (quite often no more than 8-10 people) further contributes towards interaction.

The work of study circles is governed by intentions, not by decrees and certainly not by chance. These intentions must, however, come either from the participants themselves or they must be accepted by them.

Consequently, the learning will be perceived by the participants as worthwhile. But for the work to progress according to the intentions of the participants, they require a plan for their work: "Here we stand, this we want to achieve, and that's the road we'll use."

The plan is the tool which the circle members use to secure a meaningful progress towards their goal. Such a requirement should not prevent the professional educator from finding that the learning which is worthwhile to the circle participants is also useful to him. The opposite, however, is not necessarily true.

The social interplay in study circles, combined with their guiding principles, will not only help people to overcome their isolation but also increase their knowledge and understanding of other people. They listen to the views and expressed attitudes of others. This interplay also helps members to respect different opinions and the dignity of others. It makes social and democratic learning an integral and necessary part of the learning process in those circles.

The procedures used by the study circles emphasize the equal position of all participants, including the study circle leader. The latter should not be a teacher but rather the first among equals: *primus inter pares*.

It is very difficult to describe a study circle leader's level of subject competence. No standard is known. They must, of course, have enough knowledge of the subject to make it possible for them to help the other members to progress. But big gaps between leaders and participants concerning subject knowledge often produce an undesired superior-inferior distance. This can, in turn, have a negative effect on the unity of the group and undermine the basis for circle work.

The subject level differences among the members make the competence of the leaders an important issue, not so much as far as their subject qualifications are concerned as from a pedagogical and social point of view.

The role of the leaders will, therefore, become more like that of an organizer of learning than simply a source of knowledge. Good leadership includes an interest in people, the ability to listen, to take initiative, to encourage others and to help the groups to achieve their respective aims.

To quote once more from the International Union of Bricklayers:

> Democracy requires an ongoing dialogue between leaders and the people they serve. A modern trade union also has to create mechanisms for a dialogue with its members. That is what the study circles are all about. (7)

References

1. For a more detailed discussion, see: Henry Blid. *"Education by the People - Study circles"*. Arbetarnas Bildningsforbund. Stockholm, 1990. P.O. Box 522, S-10127 Stockholm. and, Leonard P. Oliver. *"To understand is to act"*. Study circles. Seven Locks Press, Cabin John, Maryland, 1987. P.O. Box 27, Cabin John, Maryland 20818, USA.
2. One of the most important officers of the Swedish Lodge of the International Order of Good Templars, Edvard Wavrinsky, reported in 1893 of the CLSC movement in an IOGT magazine *"Julbrasan"*.
3. A Swedish researcher into the history of ideas, Maciej Zaremba, has shown that old Swedish village traditions survive in existing political organizations. The cultural heritage carried by a native Swede is seldom discussed and most often taken for granted. It might be more than a coincidence that Zaremba has a Polish background. (M. Zaremba, "Byalagetsdiskreta charm eller Folkhemmets demkratiuppfattning" In: *"Dyu skona gamla varld"* published by Sekretariatet for framtidsstudier/the Secretariat for the studies of the future. Stockholm 1987)
4. Borgstrom, Leona, and Oloffsson, Lars-Erik. "Participation in study circles and the creation of individual resources." Paper presented at the Adult Education Research Conference, Montreal, April, 1983. Department of Educational Research, Stockholm University.
5. BAC Study Circle Program: Organizing and Leading a Study Circle. *"Leader's Handbook"*. International Union of Bricklayers & Allied Craftsmen, 815 Fifteenth Street, N.W., Washington, D.C. 20005. 1987.

6. Gunnar Hirdman, "*Samtalets knost, nagot om studiecirkelns metoder*". Arbetarnas Bildningsforbunds Centralbyra, Stockholm 1945. 3:e omarbetade upplagan.
7. "*Building Union Democracy; The BAC Study Circle Program*". Executive Summary of the Final Report of the 1986 Pilot Study Circle Program. December, 1986. Report prepared by Oliver, Leonard P.

THE STUDY CIRCLE LIBRARY IN A HISTORICAL PERSPECTIVE

Lars Arvidson
Adult Education Research Group
Institute for Education and Psychology
Linkoping University, Linkoping S-581 83, Sweden

ABSTRACT

The earlier Prime Minister of Sweden, the late Olaf Palme, on one occasion named his country a "study circle democracy". He thereby indicated the important role that adult studies in small groups have played in the democratic development in Sweden and in lifelong learning. Such activities have not only expanded throughout the years but also changed in some respects. During the first decades of the 20th century, the library was the center around which these study activities circled. The study circle can, from the beginning, be regarded as a way to make the library equally accessible to every one. This author maintains that the pedagogical ideas behind the manifestation of these study circle libraries are still valid and that public libraries in the future have a great task to fulfil as a resource for adult individuals and groups in their independent pursuit of knowledge.

Background

To be able to appreciate the importance of the study circle it is necessary to draw attention to what the concept "self-education" signifies. As the term indicates, the concept means to educate oneself. However, the idea of self-education does not primarily mean individual studies but instead that members of popular movements or other collectives "take knowledge into their own hands" and create their own education. Tradition shows that self-education can take place in many forms, not least in small groups. At the end of the last century, IOGT in Sweden, for instance, gathered inspiration for such small group activities from the Chautauqua movement in the USA.

In Sweden, the library was viewed at a very early stage as the best means of realizing the idea of self-education. A movement for public libraries was already developing at the beginning of the twentieth century. (The Swedish General Library Association, still in existence, was formed in 1915.) The inspiration for this movement, among others, came from the USA and Britain. Alongside the public library, association libraries and study circle libraries were created.

Circulation libraries represented an outreaching branch of library activity. These were initiated partly by popular education associations and partly by the popular education bureaus at the universities of Lund and Uppsala. The popular education societies consisted of upper class people who eagerly supported the education of people, whereas interested university teachers took the initiative for the universities' popular education bureau. Apart from the circulation libraries, the activity that emanated from these bureaus consisted of popular lectures.

Finally, at the turn of the century, the study circle became, together with library and lecture, a more or less organized form for the practice of self-education. Education in the form of a study circle could be pursued in spite of insufficient economic means and lack of teachers. Thus, it has become the most widely spread form of popular adult education in Sweden.

The Study Circle Library and the Social Movements

From 1912, government financial support was allocated to two kinds of public libraries, municipal libraries and study circle libraries. Incidently, this financial support became the direct reason for founding the Workers' Education Association that same year.

The development of the study circle libraries has a direct relationship with that of the Swedish popular movements. The basis for their libraries was formed by the association libraries, which, in turn, could be regarded as action oriented institutions. As they were located in the same places as the association and were kept open during association meetings, they were used by thousands of people who would not otherwise have visited any library. Many people, in this way, developed an interest for reading books. In addition, this interest was strengthened by the fact that association members were themselves responsible for the library and its care.

The study circles became another pushing force in the same direction:

> An important point in the study circle programme, maybe the most important one is: "One library in every meeting place!" The task of the association is to be foci of culture in one way or another and the association members will therefore become better at this work, if they have access to a library. One bookshelf in every association as well as in every home! (1)

An essential reason for the popular movements to start building up their own libraries was that the older forms of public libraries, the parish libraries, had too limited a collection of books to fulfill the needs of the movements for their membership education. The financial support from the government gave the movements new possibilities to promote their library activities. After completing their studies on a subject, the study circles were obliged to return their books to the association library, where these sometimes comprise a special department. Those critical to the government support of study circle libraries stressed that it would be more sensible for all within a community concerned about public reading to join forces and form one single, neutral library association. Besides saving money, such an arrangement would also avoid party politics views on the choice of literature.

According to the defenders of the idea, the disappearance of the small movement libraries would cause irrevocable damage, since the task of these libraries was to spread books to as many people as possible. The municipal libraries were not as well equipped for this task as the movement libraries, despite their better resources. It was stated that in America there had long been a general understanding that the library had to go to the people if the people did not go to the library. In Sweden, the movement libraries would fulfill that progressive function. In addition, they would, in the long run, give the municipal libraries a larger audience, when they gradually became too limited for some readers. Thus, even if the movement libraries were comparatively small, their importance was considerable.

> Especially the older libraries have been of great importance in our sparely populated country, where they many times have served the public with the only good reading they have had access to. And they are all growing book collections, albeit in slow growth...Mostimportant, though, is that they are going through with the programme, "every association room also a library room," a necessary requirement for the popular associations' character as a cultural institution as well. Not only should there exist municipal libraries, but books should be accessible everywhere where people meet. (2)

The development meant a considerable expansion of the circle libraries. In 1912, municipal libraries and study circle libraries existed in about equal numbers. Nine years later, there were more than 2000 state supported circle libraries and not quite 900 municipal libraries. The number of book loans in the association circle libraries grew from 117,000 to 660,000 per year during the same time.

The library was the center of popular adult education. The study circle became, as it were, a new and more democratic way of organizing library activities, since its members had influence over the book supply. The study circle was a part of a bigger whole, where the library was the center. In 1947, government support was replaced by a subsidy towards study literature and study leaders. Consequently, the study circle libraries were gradually wound up. Within the WEA, however, a few libraries were in operation until the beginning of the 1970s.

That the study circle libraries disappeared as a phenomenon does not mean that the ideas behind their first appearance were lost. On the contrary, these ideas might be very interesting in the future "society of knowledge" or "learning society."

The Study Circle Library from a Pedagogical Point of View

First, it ought to be said that library activities in themselves were regarded as a method of self-education. Special handbooks were published, where practical instructions were given about book purchases, catalogues, book loans, etc. (3)

> At least in the country side, taking care of the library is the most important task for the study circle. The present situation is such that the numbers feel that the libraries are theirs and that care and growth of the libraries are their responsibility, their personal concern...Further, I mean that this personal interest in the library, which hundreds of thousands of members in popular associations have or will have in connection with the growth of the libraries through their contributions and their sense of responsibility, is a living force, not only in popular education as a whole but also in public library operation, which must not be jeopardized. This personal public interest is of greater importance than the most ideal centralization under the most prominent professional librarians in the most exquisite library organization. (4)

The central position of the library is based on the idea of the book as the most important source of knowledge in popular adult education. According to this idea, one meets "the great teachers of humanity" in all work of fiction, and adult individuals and groups, in pursuit of knowledge, should not settle for less.

> Books and libraries take up a central position in all popular education. It is only natural that such is the case. The book is, no doubt, the foremost means of education, and individual reading one of the most important grounds for all knowledge...Just like the library is a necessary tool for scientific research, it must also be the foundation for free and voluntary popular education. (5)

The pioneers of the Swedish study circle activities differ in their emphasis of books and libraries, from, for instance, the founder of the Danish Folk High School, N.F.S. Grundtvig, who regarded "the living work," i.e., the oral presentation and the lecture as the superior method. From a pedagogical perspective, one can apply the following viewpoints concerning the library as a form of popular adult education.

The library facilitates free, independent and joint search for knowledge.

The study circle, in a natural way, linked up with the free comradeship and social life which had developed in the regular work within the popular associations. Through the study circles, the members established closer links with the association and its meeting programme was, at the same time, raised to a higher level.

> The study circles grew from the beginning with a double task. On the one hand, they were to provide means for quenching the actual thirst for knowledge, especially among the young. As good templars, on the other

hand, they were to gather all good templars, young and old, in a highly cultivated social life around the fundamental interests in life and society. To be specific, the cultural communication was supposed to be the main point in the organization, freely chosen topics for "roundtable discussions," carefully prepared by special openers and at least somewhat reflected upon by all members through reading of literature from the study circle library. (6)

The free discourse in Plato's dialogues with Socrates as a leader was the pedagogical model for "the father of the study circle,"Oscar Olsson, when he presented his ideas at the beginning of the century. The library gave liberty, and at the same time, the organization of permanent groups in the form of study circles gave the book studies some regularity of plan. One could say that the study circle was a method to make library operations more efficient.

The library is an expression of self activity

The most important feature of self-education is initiative. Individuals and groups do not go into pre-produced patterns but decide for themselves about goals, content, methods, and time for their studies. Everyone decides about his task and pursues knowledge independently. Library work agrees well with the basic idea, even if the independent search sometimes is completed with supervision and suggestions about suitable literature.

Books give opportunities for reflection and repetition

Compared to several other methods and media, not least of which the modern mass media, the book gives opportunities to return to the material for renewed inquiry and analyses. In the library there is also access to alternative literature which can contribute to in-depth studies and to new ideas. It is to be regarded as a loss to the Swedish study circle movement that use of the library has diminished. Pre-fabricated study handbooks and modern equipment cannot, in my view, compensate for the library as a resource in the independent search for knowledge.

The book and the library represent continuity

At an early stage, the library became, as stated earlier, the center of popular adult education. Library activities were coordinated with lectures and courses as well as study circles. They gave a broad orientation, while the library made a deeper study possible. Through lectures and courses, as well as through mass media, new information of current interest can be communicated. The library in this connection offers continuity and coherence, a necessity to someone who seeks knowledge and not just information.

The study circle library was created in a special historical and social context, in close connection with the Swedish popular movements. Experiences from this work, however, indicate the possibilities in the future for the public library to act as a center for individuals and groups in their pursuit of knowledge in a deeper sense and not only as a center of information, even if the latter too is to be regarded as an important, if not necessary, task in a democratic society. Such a society needs not only well-informed but also knowledgeable citizens.

References

1. Olsson, O. *Studiecirkelbibliotek och biblioteksskotsel*. Stockholm: Oskar Eklunds bokforlag, 1916. p. 9.

2. Olsson, O. "Til Tjugarsminnet," *Bokstugan* (1922). p. 154.
3. *Op. Cit.*, Olsson, *Studiecirkelbibliotek;* and Nystrom, B. *Om lasning och bibliotek.* Stockholm: Svenska Missionsforbundets forlag, (1915).
4. Olsson, O. "Folkbildningskommitensutlatande II." *Bokstugan* (1924), p. 54 ff.
5. Tynell, F. *Folkbiblioteken i Sverige.* Stockholm: P.A. Norstedt & Soners Forlag, 1931. p. 148 ff.
6. Olsson, O. "Sammanhallning."*I Reformatorn* (1936).

Bibliography

Arvidsson, L. "Popular Education and Educational Ideology," in Ball, S. and Larsson, S., ed. *The Struggle for Democratic Education: Equality and Participation in Sweden.* New York, Philadelphia, London: Falmer Press, (1989).

Nystrom, B. *Om lasning och bibliotek.* Stockholm: Svenska Missionsforbundets forlag, 1915.

Olsson, O. "Folkbildningskommitensutlatande II." *Bokstugan,* (1924).

Olsson, O. "Sammanhallning."*I Reformatorn* (1936).

Olsson, O. *Studiecirkelbibliotek och biblioteksskotsel.* Stockholm: Oskar Eklunds bokforlag, 1916.

Olsson, O. "Til Tjugarsminnet," *Bokstugan* (1922).

Tynell, F. *Folkbiblioteken i Sverige.* Stockholm: P.A. Norstedt & Soners Forlag, (1931).

U.S. GOVERNMENT SUPPORT FOR CONTINUING PROFESSIONAL EDUCATION FOR LIBRARIANS

Anne J. Mathews
Director
Office of Library Programmes
Office of Educational Research and Improvement
U.S. Department of Education, 555 New Jersey Ave., NW,
Washington DC 20208-1430, USA
and
Thomas C. Phelps
Senior Programme Officer for Libraries and
Assistant Director of the Division of General Programmes
National Endowment for the Humanities, Old Post Office,
1100 Pennsylvania Ave., NW, Washington DC 20506, USA

ABSTRACT

This paper describes two major programmes supported by the U.S. government through U.S. Congressional appropriations, administered by the U.S. Department of Education. A third programme supported by a federal funding source, the National Endowment for the Humanities, is also described.

The United States government supports the professional education of librarians through a variety of legislated programmes. However, only two of these federal funding programmes are targeted specifically for library and information science education. Other federal funding sources support grant programmes for purposes other than libraries, but libraries and librarians may participate in them. This paper describes the two major programmes supported by U.S. Congressional appropriations and administered by the U.S. Department of Education, as well as one other federal funding source, the National Endowment for the Humanities.

U.S. Department of Education

The U.S. Department of Education oversees the funding programmes legislated exclusively for libraries. Within the Department, the Office of Library Programmes of the Office of Educational Research and Improvement carries out this responsibility, administering both the Library Services and Construction Act (LSCA) and Title II of the Higher Education Act (HEA II).

LSCA supports the following:

- public library services, especially to underserved populations;
- public library construction;
- networking and interlibrary cooperation;
- library services to Indian tribes, Alaskan native villages, and Hawaiian natives; and,
- adult literacy programmes in public libraries.

HEA, Title II, provides funding to:

- strengthen major research libraries;
- promote and develop exemplary uses of technology for resource sharing and networking in colleges and universities;

- promote research and demonstration projects on library and information science issues; and,
- advance the education of librarians through fellowships and training institutes.

While library career training is a specific programme under HEA II, professional education for librarians is not a designated purpose of LSCA. However, training is an eligible expenditure under the LSCA if it supports the purposes described in the legislation.

Higher Education Act

The purpose of the Library Career Training Programme, Title II-B of the Higher Education Act, is to provide grants to institutions of higher education, and other library organizations or agencies, to train persons in librarianship; and to establish, develop, and expand programmes of library and information science. There are two parts to the programmes, Fellowships and Institutes. Since 1966, the Library Career Training Programme has provided funds totaling $51,514,070 to assist 20,766 persons in training and retraining for library service.

The Fellowship programme was established to alleviate the critical shortage of professionally trained librarians and the serious shortage of graduate library school faculty. In the early 1970s, the programme was redirected to recruit minorities to the profession, to facilitate upward mobility of women and minorities, and to enhance service to the underserved. To date, $32,141,259 has been spent to support 16 bachelor's, 53 associate, 77 traineeships, 2,808 master's, 248 post-master's, and 1,101 doctoral fellowships.

The Institute programme was established to provide long- and short-term training and retraining opportunities for librarians, media specialists, information scientists, and persons desiring to enter these professions. To date a total of 16,453 people have been trained through 435 institutes funded through grants totaling $19,372,811. Many of the institutes have given experienced practitioners the opportunity to update and advance their skills in a given area of librarianship.

The Library Career Training Programme addresses the training needs of the library community in order to prepare librarians for service where shortages of qualified professionals are critical. The current priorities for the programme include six broad categories.

They are:

1. To train or retrain library personnel in areas of library specialization where there are currently shortages, such as school media, children's services, young adult services, science reference, and cataloging;
2. To train or retrain library personnel in new techniques of information acquisition, transfer, and communication technology;
3. To increase excellence in library leadership through advanced training in library management;
4. To increase excellence in library education by encouraging study in librarianship and related fields at the doctoral level;
5. To provide advanced training in the development, structure, and management of new library organizational formats, such as networks, consortia, and information utilities; and,
6. To train or retrain library personnel to serve the information needs of the elderly, the illiterate, the disadvantaged, or residents of rural America;

Promulgating annual priorities allows the programme flexibility by responding to the changing needs of library and information science training.

There is a strong interest in library education in the United States. For fiscal year 1989, approximately $400,000 was appropriated by the U.S. Congress for training activities under HEA Title II-B. However, there were requests for 144 fellowships totalling $1,509,500, and 19 applications for institutes totalling $949,022, a combined total of $2,458,522, or six times more dollars than were available.

All of the applications were reviewed by peer reviewers who volunteered their time. Each application was reviewed by three persons, and the three scores were averaged. All scores were ranked from highest to lowest.

When the results were tallied, 12 master's fellowships were funded for $129,600, and 10 doctoral fellowships were funded for $148,000, a total of $277,600. In addition, four institutes were funded for a total of $122,187. Following are four examples of the 22 fellowships that were funded in fiscal year 1989:

- Catholic University of America, School of Library and Information Science (Washington, DC) received a $10,800 grant to provide a fellowship for the Master's degree, training a student with a background in basic life sciences, specializing in science reference with an emphasis on biomedical sciences, an area of library specialization where there is a current shortage.
- Indiana University, School of Library and Information Science, used a $29,600 grant to support two doctoral fellowships, one specializing in implementing and evaluating network technologies as they apply to a major research library, a major institution providing library and information science education, or networking for youth between school media centers and public libraries. The second fellowship specialized in public or school library services to handicapped youth and research methods which lead to new knowledge concerning the development, implementation, and evaluation of such services.
- Louisiana State University, School of Library and Information Science was granted $10,800 to offer a fellowship for Master's level training in current forms of bibliographic control and fundamentals of database organization and management for application in a research, special, or public library.
- University of Wisconsin-Milwaukee, School of Library and Information Science directed its $10,000 grant to support one Master's level fellowship toward preparing a student for a career in library services for children and young adults in an urban public library. The programme of study included opportunities to work with practicing librarians, taking part in established service programmes for children and young adults.

In fiscal year 1989, four institutes were funded. They were:

- Rosary College, School of Library and Information Science (Illinois) received a $28,280 grant for a one-week institute to introduce 75 librarians to the issues, problems, and opportunities facing the management of networks.
- Indiana University, Graduate School of Library and Information Studies used its $28,172 grant to sponsor a two-phase summer institute, a three-day forum for library educators and a two-week session for school library media specialists and public librarians, on library services to youth with handicaps.
- The Board of Education, Prince George's County (Maryland) applied its $31,815 grant to train 35 traditional school librarians as information specialists, emphasizing acquiring and managing information and technological change.
- Oklahoma State Department of Education, with its $33,420 grant, supported an eight-day institute to train 35 school library media specialists and five library educators in the application of microcomputers in library management, including the conversion of bibliographic records to machine-readable form and micro-based resource sharing.

The U.S. Department of Education training monies are awarded to academic institutions and libraries. Persons interested in receiving a fellowship or attending an institute apply directly to the grantee institution, which has the sole responsibility for selecting recipients or participants.

Institutions receiving grants do not require fellows or institute participants to pay tuition and fees. Stipends for fellowship recipients range from $5,400 for master's to $7,400 for post-master's and doctoral level study. For fellowships, the institution receives an amount equal to the stipend to cover the cost of training and the cost of waiving tuition and fees.

Institutes' costs vary according to the length of time and number of participants. Other considerations include grantees' costs for personnel, support services, and other administrative costs.

Library Services and Construction Act (LSCA)

Since its creation in 1965, LSCA has allowed for the professional education of librarians. Most of the funding appropriated under LSCA goes directly to the States to support improvements in public library services, and resource sharing and networking. This programme allocates money to the States based on their populations, and the programme is administered by the State library agencies to meet the needs identified in their long-range plans. As long as education programmes for librarians meet the State's identified needs as well as the criteria established for LSCA, e.g., for Title III, the programme must support more than one type of library, they can be supported in whole or in part by LSCA funds.

Examples of professional education programmes for librarians supported by LSCA grants (Titles 1 and III) to the states include:

- In Massachusetts, a $25,000 Title I grant supported two regional workshops aimed at strengthening management skills of public library directors and supervisors. Training included budgeting, team-building, performance appraisal, staff development, decision-making, time management, communication skills, and positive assertive management.
- A large public library in Florida, with a grant of $27,075, trained its online searchers and key reference personnel in the use of various online systems such as DIALOG and BRS.
- The Delaware Division of Libraries used $14,800 of their grant funds to hold two workshops on reference skills and business resources, and collection development. Thirty-six librarians attended the workshops, which were offered for certification/re-certification credit.
- A library cooperative in California received an $8,070 grant to help sponsor the Third Pacific Rim Conference on Children's Literature in cooperation with the California Association of the Teachers of English, the Children's Film and Television Center of America, and the Graduate School of Library and Information Science at the University of California at Los Angeles. The conference brought together distinguished authors, librarians, and other persons working in the field of children's literature from North and South America, Asia, and the Pacific.
- A cooperative in Virginia, using a $5,400 Title III grant, conducted workshops on four topics of interest to local librarians: public relations for library personnel, gift materials, cataloging of audiovisual materials, and business reference sources and services. A total of 188 people from 18 libraries, including school, public, academic, hospital, and special libraries, participated.
- In New York, a local library council received a $16,438 grant to develop a training programme for coordinated collection development in allied health sciences and to produce a regional plan for all types of libraries. The librarians were trained in collection assessment procedures and techniques, analysis of user needs, identification of interlibrary loan patterns, and interpretation of the data from OCLC

holdings tapes for participating libraries. Participants learned to collect and develop more meaningful statistical data on the usage of their collections.

- The California State Library used Title I funds to further its minority recruitment programme for librarians. "Transition into Management" was a one-week middle management seminar for ethnic librarians who had been in the field at least three years and desired to move into middle management positions. Out of 30 applicants, 10 participants, 5 Black, 2 Asian, 2 Hispanic, and one American Indian, were selected to attend the University of California at Los Angeles seminar which covered a number of issues, including interviewing and resume writing, and offered opportunities to network with successful ethnic librarians.

Title IV of LSCA is a grant programme directly administered by the Office of Library Programmes. It is devoted to supporting library services for Indian Tribes and Hawaiian Natives. A number of tribes have used all or part of their grants to train personnel. For example:

- The Three Affiliated Tribes of New Town, North Dakota, selected a tribal member to attend Clarion University of Pennsylvania to work on a Master's degree in rural librarianship. Having a qualified librarian with special training in library services to rural areas will improve services to the Tribes. Since the librarian is not a full-time student, the Tribes will use grants for several years to support the education programme.
- The Pilot Point, Alaska, Native Village used part of its grant to send its librarian to the Village Library Workshop. The workshop included such topics as basic library skills, library grant management, children's services, and policies and procedures for participation in the State interlibrary system. The Alaska State Library has sponsored the Village Library Workshop for several years to train personnel serving Alaskan Native Villages.

As more libraries participate in literacy activities, there is an increasing need to train more librarians to conduct literacy programmes. Under LSCA title VI, Library Literacy Programme, many State libraries are training librarians to develop and implement literacy programmes or provide particular literacy services. At the local level, public libraries are training librarians to carry out specific literacy projects. For example, some librarians are trained as tutors, others as tutor trainers, and still others in developing collections of materials for adult new readers. Two examples of how Title VI funds are used for training follow:

- One urban public library system in Florida, with a grant of $24,945, is training library staff and volunteers from 12 communities to provide literacy services that will be accessible to nonreaders and new readers.
- The Oregon State Library used its grant of $25,000 to solicit applications from local libraries, select the 50 with the greatest need to establish literacy programmes, and train the librarians and project coordinators in various topics related to establishing and managing local library literacy programmes.

Other U.S. Government Funding Sources

While the Library Services and Construction Act and Higher Education Act, Title II, programmes are specifically targeted to libraries and the training of librarians, other government agencies also provide support for the training and education of professional librarians. The National Science Foundation, the Smithsonian Institution, National Institute for Museum Services, National Aeronautics and Space Administration, National Trust for Historic Preservation, National Libraries of Agriculture and Medicine, and National Endowment for the Humanities are a few such agencies.

Appropriations from the U.S. Congress to three agencies include provisions for training librarians for collection building and maintenance in particular subject areas, for access, and methods of interpretation of collection works; and for strategies for research through particular resources. Such assistance allows the agencies to fulfill specific mandates or goals. Therefore, workshops, seminars, sessions, and conferences are often a part of continuing education supported by the U.S. government. A closer look at the National Endowment for the Humanities demonstrates the types of training programmes funded by such agencies.

National Endowment for The Humanities

To promote progress and scholarship in the humanities in the United States, Congress enacted the National Foundation on the Arts and Humanities Act of 1963 which established the National Endowment for the Humanities (NEH) as an independent grant-making agency of the federal government to support research, education, and public programmes in the humanities. The term *humanities* includes the study of history; philosophy; languages; linguistics; literature; archaeology; jurisprudence; the history, theory, and criticism of the arts; ethics; comparative religion; and those aspects of the social sciences that employ historical or philosophical approaches or methodologies.

While not specifically included in the Act of 1965, nor viewed as a discipline in the humanities, librarianship is a field of study useful to the humanities. Libraries collect, preserve, and maintain a major portion of the recorded cultural heritage and make it available to scholars and the general public. Because libraries and librarians are important to research and education in the humanities, they are important to the NEH. Therefore, the NEH makes grants to libraries for access to collections, for public education in the humanities using library collections, and for the preservation of collections.

Libraries play a primary role in research and public education in the disciplines of the humanities. Librarians (public, school, academic, and research) organize, house, care for, and assist in accessing materials, and facilitate the interpretation of material for the general public and for students. Therefore, it is some concern to NEH that librarians receive training and continuing education so that they are knowledgeable and skillful, especially as they perform work relating to NEH's mandate.

NEH makes direct grants to a variety of agencies and institutions, including libraries. Within grant requests, funds may be budgeted for the continuing education of librarians. Grants that will advance an understanding or appreciation of the humanities have been made to agencies such as the American Library Association, the Northeast Document Conservation Center; or to universities such as Johns Hopkins or Columbia; or to libraries such as the New York Public or the Newberry. Each of these grants has included training librarians to facilitate access to, interpretation of or the preservation and conservation of collections of materials vital for the study and appreciation of subjects, topics, themes and disciplines of the humanities.

For example, the American Library Association received a large grant to support a workshop for librarians and scholars in humanities disciplines from historically black colleges and universities. The work assisted participants in planning and conducting educational programmes for the general public on themes and topics about black American literature and culture. During the workshop, collections were analyzed and topics, ranging from Langston Hughes to the Harlem Renaissance, derived from them.

Both librarians and scholars were trained to interpret collections for the general public to help them better understand American culture.

Other examples of support for continuing education for librarians include grants for specific kinds of training. With NEH support the following projects were undertaken:

- The American Antiquarian Society established an endowment for a professional conservator and an intern for the conservation of the Society's collection and for training other selected individuals in the conservational craft.

- Columbia University set up a programme for graduate instruction in conservation and preservation for librarians already advanced in their careers.
- The Northwest Document Conservation Center developed field services to provide information and training for librarians, archivists, and curators on-site at their own institutions.

While the NEH is not in the business of directly supporting continuing education for librarians, through indirect measures it offers support to libraries and institutions of higher education for the continuing education process.

Conclusion

The scope and variety of the training opportunities described in this article reflect the ever-changing and diverse nature of the library profession itself. Developments in technology, sophisticated approaches to information management, increased demands of researchers, and delivery of information through international databases challenge educators and practitioners to prepare, train, update, and retool for more effective delivery of library services in the 21st century. In the United States, the federal government plays a major role in support of these needs.

LIBRARY SCHOOLS OFFERING TRAINING COURSES IN SERVICES TO THE PRINT-HANDICAPPED: AN UPDATE

Bruce E. Massis
Director of Library Services
The Jewish Guild for The Blind
15 W. 65 Street, New York, NY 10023, USA;
and
Member, IFLA/SLB Working Group on Continuing Education

Introduction

In June 1976, a survey of graduate library schools in the U.S. and Canada was conducted by the National Library Service for the Blind and Physically Handicapped in an effort to identify existing training programmes in library services to the print-handicapped. The survey results were analyzed and appeared as a published article in the *Journal of Education for Librarianship* in that year.

In 1989, an update of that survey was undertaken by a working group within the IFLA Section of Libraries for the Blind in order to discover the progress (or lack thereof) in this area of professional education. The question nagging both sets of researchers was:

> "Are the library schools training our librarians adequately to work within the field of services to disabled readers?"

Historically, the lack of proper professional training in this area has been cause for librarians to face their jobs with a mixture of professional savvy and sheer instinct. Paul Thiele, head of the Crane Memorial Library in Vancouver, B.C. has written:

> ...It is a tremendous compliment to the curators of these collections who often had to develop their own home-made processes of classification, cataloging and lending without the benefit of formal training...

During its first decade, members of the IFLA Section of Libraries for the Blind and others have produced a small library of articles and publications designed to support international cooperation and to be used as reference materials for librarians working in and maintaining libraries serving the disabled, the blind and visually handicapped. Thus for librarians already working in the field, their work has become easier. But what of the formal training required by new library school graduates? Have they received their training before entering the field? We will look at the figures to find out.

As the number of library schools in North America decreases (67 in 1976, but only 53 in 1989), the opportunities for such formal training decrease as well, or does it? First, let us look at the surveys themselves.

Questionnaire on Library Service Courses - Service to Handicapped Groups

Do you currently offer any courses especially concerned with library services to the handicapped?
Yes_____ No_____ If yes, please list the course titles.

If you do not offer such courses, do you plan to do so in the near future?
Yes_____ No____

Have you recently conducted workshops, seminars. etc. on library services to the handicapped?
Yes_____ No_____
If yes, were they held? on campus_________off campus__________

Are you willing to conduct or sponsor workshops on library services to the handicapped?
Yes______ No_______

Do you wish to receive a copy of the list we will compile from the result of this questionnaire?
Yes______ No______

Name____
School____
Address____

The NLS Survey Results

			Number of Schools	% of Schools
I	**Formal Courses**			
	a. Presently offered		12	20
	b. Planned in future		1	2
	c. Not planned		43	73
		Total	56	95
II	**Special Institutes**		1	2
III	**Specialized Independent Study**		2	3
		Total (I, II, III)	59	100
IV	**Workshops and/or Seminars**			
	a. Presently or recently conducted		9	15
	b. None conducted but willing to sponsor		45	76
	c. Not interested in sponsoring		5	9
		Total	59	100
V	**Response to Questionnaire**		59	88
VI	**No Response to Questionnaire**		8	12
VII	**Total Number of Schools**		67	100

The 1989 IFLA/SLB Working Group on Continuing Education Survey

Do you presently offer any courses on library services to the handicapped?
Yes_____ No_____

Do you offer any short-programmes (workshops, seminars, etc.) on library services to the disabled?
Yes_____ No_____

Do you plan to offer such a course or short programme in the future?
Yes_____ No_____

Are you aware of resources available in your city to assist in the preparation of such a course or short programme?
Yes_____ No_____

Would you be willing to offer such a course or short programme off-campus in an accredited agency serving the disabled?
Yes_____ No_____

Would you be willing to invite a librarian presently serving the disabled to speak to your students on such services to create an awareness of these activities?
Yes_____ No_____

Name____
School____
Address____

Breakdown of 1989 Survey Results

Table I: **Library schools in North America presently offering an accredited course of training in services to the disabled**

1) Catholic University of America (Washington, D.C.)
2) Clarion University (Pennsylvania)
3) Long Island University - C.W. Post (New York)
4) University of Maryland
5) St. John's University (New York)
6) Southern Connecticut State University
7) Atlanta University (Georgia)
8) Florida State University
9) Louisiana State University
10) University of South Carolina
11) Emporia State University (Kansas)
12) Indiana University
13) Kent State University (Ohio)
14) University of Missouri, Columbia
15) University of Washington

Table II: Library schools in North America presently offering a short course, seminar or workshop on services to the disabled

1) St. John's University (New York)
2) Florida State University.
3) Louisiana State University
4) University of North Carolina
5) University of Tennessee, Knoxville
6) Emporia State University (Kansas)
7) Kent State University (Ohio)
8) University of Michigan
9) Northern Illinois University
10) University of Wisconsin, Madison
11) Texas Woman's University
12) Brigham Young University (Utah)
13) University of Washington
14) University of British Columbia

Table III: Library schools offering information on library services to the disabled as part of a general survey course

1) State University of New York, Buffalo
2) University of Alabama
3) University of Illinois
4) University of Iowa
5) University of Michigan
6) Texas Woman's University
7) University of Hawaii
8) Dalhousie University (Halifax, Nova Scotia)

Table IV: Library schools not currently offering such a course but planning to do so in the future

1) North Carolina University
2) Rosary College (Illinois)
3) University of Wisconsin, Milwaukee
4) University of Hawaii
5) San Jose State University
6) University of Toronto

In answer to the survey's fourth question, "Are you aware of resources available in your city to assist in the preparation of such a course or short programme?" the response was as follows:

YES:	40
NO:	8

In answer to the survey's fifth question, "Would you be willing to offer such a course or short programme in an off-campus agency serving the disabled?" the response was as follows:

YES:	37
NO:	8
PERHAPS:	3

In answer to the survey's final question, "Would you be willing to invite a librarian presently serving the disabled to speak to your students in such services to create an awareness?" the response was as follows:

YES:	41
NO:	5
PERHAPS:	2

Statistical Comparison of 1976 and 1989 Surveys

		1976	1986
I	**Formal Courses**		
	a. Presently offered	12	15
	b. Planned in future	1	6
	c. Not planned	43	27
II	**Workshops/Seminars**		
	a. Presently or recently conducted	9	14
	b. None conducted but willing to sponsor	45	40
	c. Not interested in sponsoring	5	8
III	**Total Number of Schools**	67	53
IV	**Response to Questionnaire**	59	48
V	**No Response to Questionnaire**	85	--

In the main categories, it may be noted that across-the-board percentages have increased during the past decade. "Formal Courses" offered increased 11%, "Workshops," doubled its category and increased 15%, "The willingness to sponsor" increased 7% and the general survey responses by the library schools increased 3%. This a good sign, but it is by no means terribly optimistic. Though certainly more widespread and acceptable, training in service to the disabled remains an area where much growth is needed. There are still 24 library schools in North America with no training in this area at all and 38 of 53 schools who do not offer a "formal" course.

It also may be noted that certain library schools who offered a "formal" course in 1976 no longer do so, including:

- Columbia University (New York
- Drexel University (Pennsylvania)
- University of Denver (Colorado)
- University of Minnesota
- Wayne State University (Michigan)
- Emory University (Georgia) - the Library School has been closed.

Conversely, the 1989 survey contains a number of schools who have added a "formal" course to their curricula in the past decade. Those schools include:
- Clarion University
- Long Island University (C. W. Post)
- Southern Connecticut State University
- Atlanta University
- Florida State University
- Louisiana State University
- Indiana University
- Kent State University
- University of Missouri, Columbia

A hearty "bravo" to these schools. Through the continuing efforts of the IFLA/SLB working group and others, there may, hopefully, be more on this list in the future.

Thus, the next action to be taken by the working group is clear: to analyze the course outlines of existing courses in services to the print-handicapped. From that information, a "sample" course will be constructed from all the most relevant aspects of present courses and used as an aid by the IFLA/SLB working group to convince schools not presently offering such a course to do so.

The profession must face the fact that as greater numbers of the population age, the need for trained librarians in the field of service to those patrons requiring reading matter in alternative formats will be greater than ever. It is up to the library schools, a librarian's first encounter with various types of library service, to recognize this pressing need.

Bibliography

Gibson, Merrillyn C., "Preparing Librarians to Serve Handicapped Individuals," *Journal of Education for Librarianship*, 1976, pp. 121-130.

Lucas, Linda, "Education for Work With Disabled and Institutionalized Persons," Winter, 983, pp. 207-223.

M[c]Queen, Lorraine, "Continuing Education for Library professionals Serving Blind persons," August, 1986, IFLA Expert Meeting f/t Blind Paper, Tokyo, Japan.

Thiele, Paul E., "Continuing Education and Training in Library Service for Blind Persons," August, 1985, IFLA Expert meeting f/t Blind Paper, Washington, D.C.

GENERAL PROGRAMME FOR EDUCATION IN AGRICULTURAL INFORMATION (PROGEFIA)

Michel J. Menou
Consultant in Information Management Systems
129 Avenue Paul Vaillant-Couturier
Gentilly 96250, France

The General Programme for Education in Agricultural Information, PROGEFIA, is a long term strategy for the transformation of educational programmes in agricultural information offered in the developing countries and their progressive transfer to the educational institutions of these countries. It proposes a coherent series of modular reference programmes in response to the needs of the various categories of personnel which contribute to agricultural information activities, in whatever capacity or level (managers, staff of the technical and administrative units, staff of the information units, trainers). This conceptual framework should in the first place allow for the establishment or strengthening of synergies between the participating organizations facilitating the combination of their efforts, on a voluntary basis. It should further provide guidance in the consolidation and development of programmes which could be more easily transferred.

The improvement of the conditions of the rural communities which form the largest portion of the population in developing countries requires that agricultural institutions increase their productivity. In order to do so, they have to establish effective mechanisms for the management of agricultural information, both internally and at the national and international levels. This calls in turn for the availability of qualified personnel at all levels, which is itself dependent from the provision of appropriate basic and continuing education.

For the time being, and probably for quite some time in the future, the agricultural institutions in the developing countries maintain only a rather loose set of formal information units. This implies that a good deal of the functions of such units should be subsidiarily secured by technical or administrative staff. The measures for the reduction of government spending also result in the fact that this type of personnel which is readily available has to be assigned to the information units as they are established. On the other hand, even if the information infrastructure is well developed, it could effectively operate only when the entire structure of the organizations and their staff are able to perform those information tasks which are in any case part of their duties. As a matter of fact, in the information society which is taking shape in all continents as the information and communication technologies become more and more effective and economical, all organizations face the necessity of intelligently mobilizing all their information resources in support of their strategic objectives. This goes far beyond the more or less coherent juxtaposition of specialized information services each responsible for a special type of processing of specific kinds of information, even with modern techniques. The design of PROGEFIA relies therefore upon the concept of information resources management. All the roles and functions in the communication cycle, from the production of all kinds of documents to the management of national agricultural information systems (1), are taken into consideration, in view of their functional and sociocultural interdependence.

Most of the personnel involved in such activities have received a basic education in an agriculture or administration related discipline, which, except for scarce courses in specialized bibliography, makes little or no room for a specific preparation to information activities. A very small portion of the staff had its basic education in library/information science programmes which are mostly general in nature and seldom include subject specialization courses. As a matter of fact, SLIS University of Pittsburgh is currently the only institution in the world to offer an agricultural information specialist programme.

Specialized education and training is thus primarily dependent at the moment from continuing education programmes, through either short courses, in-service training or training components of projects. The present provision of such educational opportunities, though it may be satisfactory in specific and local instances, is inconsistent and does not properly respond to the needs of the people nor to the manpower development requirements of the national agricultural information systems (2). Numerous organizations, generally based in industrialized countries, working independently, offer specialized short courses, mostly one by one, which increases the relative cost of their preparation and implementation.

This scatter makes information difficult to access and hinders the pooling of efforts. Courses repeat rather than complement each other. They do not grant the diplomas which are required in order to secure satisfactory working conditions and career prospects. Most of these training activities are not allocated regular funding. Irrespective of their type, they are constrained by the scarcity of qualified educators, the lack of suitable training and learning material, the scarcity of excellence centres which could host practical training sessions in conditions close enough to real life ones. The available resources do not allow for the full training of a sufficient number of persons during a given time span so that a critical mass could be achieved. Trained people remain isolated in a non-supportive environment without having access to a real follow-up and technical backstopping, further continuing education or technical information. For many reasons, staff turn over is high and is to some extent encouraged by training.

Even in a long term perspective, no organization could possibly mobilize alone the resources which are required in order to alleviate these limitations. Thus the need for combining the efforts of all those concerned. In an informal meeting held in Rome at the invitation of the Technical Centre for Agricultural and Rural Cooperation (CTA), an organization of the ACP/EEC Lomé Convention, representatives of 11 national and international organizations active in training for agricultural information suggested that a programme be prepared with a view to facilitating the harmonization of these activities, improving their adequation for the needs and to better attract regular funding (3). CTA commissioned the preparation of an outline of such a programme (4), which was distributed in the summer of 1989 to some 300 specialists and organizations in 80 countries, together with a structured questionnaire allowing them to express their own views as to the desirable features of the programme. On the basis of the responses received (about 30 percent), a new version of the programme proposal as well as a draft action plan for its implementation during 1990-1992 (5) were prepared and submitted to an Experts consultation which met in Libreville, Gabon, on December 18-21, 1989. On this occasion, the general programme and action plan were approved with a few slight amendments.

The general programme is structured into eight major programmes:

1. Implementation of specialized training courses in agricultural information;
2. Support to the establishment of regular academic programmes in agricultural information (both in library/information science and in agriculture or administration related schools);
3. Training of trainers in agricultural information;
4. Production of teaching and learning material for education in agricultural information;
5. Joint support services (standing inventories of educational resources, technical

bulletin for current awareness of trainees and former trainees, clearinghouse support to individual course assignments);

6. Strengthening of educational infrastructures (educational institutions, excellence centres for practical training, support to local continuing education programmes);
7. Development of the general programme and instructional methods in agricultural information (continuing development of the model programmes for specialized education and the training of trainers programmes, development of instruction methods and tools toward the systematic use of distance education methods based upon modern information technology, study of statutory constraints); and,
8. Overall coordination of PROGEFIA.

In its present version, PROGEFIA identifies nine programmes, which may themselves be more specifically geared in particular functions through 16 tracks. All have a modular structure which should facilitate both the implementation of the courses and the sharing of resources among participating organizations. PROGEFIA further looks for the provision of a complete professional preparation through the implementation of adapted series of modules, as opposed to the scattered and loosely related single courses which are currently offered. All programmes include a specific subset of agriculture related courses in order to provide a global perspective of the field and/or complement the subject knowledge of people with a basic education in those disciplines or provide a minimum subject background to those with a basic education in other areas.

It is expected that the implementation of PROGEFIA will primarily be carried out at a regional level. The delimitation of the regions will be at the discretion of the participating organizations. The establishment of the regional coordination is among the priority tasks of the 1990-1992 action plan. In each region, the participating organizations will set up one or several consortiums in order to organize their cooperation toward the implementation of those parts of PROGEFIA which they will have selected for each period as the priority ones in view of the local conditions and resources available. A general coordination will be responsible for the liaison among the regional coordinations, the complementarity between their activities, the sharing of experience and resources, the monitoring and the development of PROGEFIA as a whole and the implementation of activities or provision of the joint support services which concern all the regions. It will be entrusted to a liaison committee, a smaller steering committee and a permanent secretariat. CTA has accepted the responsibility of the Permanent secretariat for 1990-1992 on the assumption that other interested organizations will also contribute to the support of the general coordination.

During the period covered by the Action plan 1990-1992, which will serve as a transition phase, four main objectives will be considered:

1. The joint implementation of short specialization courses, taking advantage of the regular or already planned activities, with a view to undertaking small scale experiments of the training methods recommended in PROGEFIA. To this end, the present courses would need to be strengthened by appropriate complements, as feasible within the available resources, concerning the provision of teaching material, the follow-up activities, the linking of the courses into more comprehensive programmes and possibly their recognition. This shall also lead to the setting up of the joint support services as required by the on-going pilot activities and the development of the other components of PROGEFIA;
2. The beginning of the training or trainers activities in conjunction with the planned specialization courses;
3. The establishment of regular procedures of cooperation and monitoring of the activities among the participating organizations; and,

4. The continuation of the studies and discussions with potential participants for the definition of the contents and modalities of the general programme as well as for its organization.

The following example may illustrate the kind of transformation PROGEFIA is aiming at. In a given region, several organizations may have each planned one or several courses for the personnel of agricultural institutions, which roughly relate to the same general topic, e.g., the use of micro-computers for the handling of agricultural data bases. In the framework of PROGEFIA, these courses could be adjusted and combined in order to form a series allowing for a more comprehensive and effective coverage of the subject, e.g., introduction to micro-computers, establishment and maintenance of agricultural data bases, management of agricultural data bases.

Once the courses have been so redesigned, they can be scheduled over a period of 18 to 14 months. The participants selected for the courses could be asked to undertake individual preparatory work before the first course. At the end of the latter they would be assigned adapted application working their own institution for a period of several months, which would be monitored at distance by members of the instructors team. Admission in the second course will be subject to the evaluation of achievements during the first one and the application period. The same scheme will apply for the further courses. Basic teaching and learning material will be distributed and participants will receive an appropriate technical bulletin for keeping abreast with developments in the field during the entire cycle.

Participation in the courses will also be conditioned by the availability of related resources in the respective organizations. To a limited extent, ad hoc support may be granted in order to overcome unexpected material limitations which could prevent the proper implementation of the individual assignments.

It is further anticipated that the implementation of such series of courses will systematically be undertaken with the participation of teachers from the specialized schools in the region so that they could progressively take over the responsibility of their organization.

Through such an integrated and cooperative approach, instead of providing a relatively superficial training to sixty people in three distinct groups and leaving them without operational skills, support and prospects, 20 people could acquire a full command of the considered technique and effectively put it into practice in their institutions.

For the long term, when conditions will so permit, it seems that distance education methods would be the most appropriate in order to provide this type of training through a combination of series of residential sessions for basic theoretical and practical instruction, sessions of supervised practical work, periods of individual or group practical application work directly applied in the home institutions, personal studies and work. Such an approach is the only one which allows for keeping the staff on duty during its training and immediately translating the acquired skills and knowledge into the professional practice. It is also the only one which fits with the unavoidable dispersal of the qualified trainers among a large number of institutions in many different countries. Distance education methods can only be effective if the educational programmes have been designed for this purpose, if the required investments took place, especially for the production of comprehensive sets of teaching and learning material, if the trainers have been prepared for using them, if the employing organizations are committed to the implementation of the training, if the participants are motivated and benefit from conditions facilitating their individual work, and finally if reliable communications channels are available. Many of these conditions do not yet exist in the developing countries. However, the present deficiencies in the communication facilities are not, as often advocated, a sufficient reason for discarding the application of distance education, since they may become available much sooner than the educational resources, unless the noteworthy investments which the latter require are initiated without delay.

A tentative programme for the training of trainers has been designed. It would include three main areas of studies: general courses (to complement and consolidate basic knowledge, especially for expression and communication techniques), agricultural information systems, and teaching methods. As part of the action plan 1990-1992, it is anticipated to hold a seminar in order to complete the design of the programme, to survey potential instructors and establish a standing directory in cooperation with interested organizations and to plan for the further implementation of the programme. If available resources so permit, courses shall begin in the same period.

References

1. Borko, M. & Menou, M.J. *Index of Information Utilization Potential (I.U.P.)* Paris: Unesco, 1983, 114 p. Serial N. PGI-1-83/WS/29.
2. Martinelli, M.T. Ronchi. *"Education and Training for the Agricultural Information Professions". Quarterly Bulletin of the International Association of Agricultural Librarians and Documentalists* 28 (3): 126-151. (1983).
3. CTA. Reunion sur la cooperation internationale en matière de formation des bibliothécaires et documentalistes agricoles, Rome, Italie, 17 juin 1989; Rapport et recommendations. Ede, Netherlands, CTA, July 1988, 9 p., Ref: 729-TN/ES.
4. Menou, M.J. and T. Niang, General programme for education in agricultural information, PROGEFIA. Document N. 1. Outline of PROGEFIA. Ede, Netherlands, CTA, July 1989, 57 p. Available from CTA - Technical Centre for Agricultural and Rural Cooperation, P. O. Box 380, 6700 AJ Wageningen, Netherlands.
5. Menou, M.J. and Niang, T. General Programme for education in agricultural information, PROGEFIA. Document N. 3. Draft programme and action plan 1990-1992 for the implementation of PROGEfIA. Ede, Netherlands, CTA, December 1989, 74 p. Available from CTA - Technical Centre for Agricultural and Rural Cooperation, P. O. Box 380, 6700 AJ Wageningen, Netherlands.

THE LIBRARY AND INFORMATION SCIENCE PROFESSION: ANTICIPATION OF FUTURE IMPORTANT CHANGES LOCAL, NATIONAL, INTERNATIONAL

Martha Boaz
Dean Emeritus and Professor Emeritus
Graduate School of Library Science
University of Southern California, USA

ABSTRACT

Library and information science professionals should be aware that dramatic changes are essential if the profession is to grow. The author outlines necessary changes in educational requirements, suggests continuing education activities, promotes a research agenda, details possible funding sources, and calls for expanded leadership.

The ever-growing flood of information, its complexity and the use of sophisticated technology, have brought about dramatic changes in the library and information science profession, its ability to handle change and ways to improve its services. This is true on local, national, and international levels. Involved in this are the educational preparation of practitioners in the field, the services of working professionals, and change in both their preparation and their practice.

This is an information age. World knowledge is vital to world work and library and information scientists have a serious responsibility to provide this knowledge. Changes in their educational preparation and their professional services, after graduation, are essential to world welfare. The material that follows addresses some of these issues and emphasizes the importance of international world cooperation, among library and information professionals.

The field of information management has undergone dramatic changes since 1930 and the world is now in an information age, an age in which sophisticated information and communication of this information influences decisions which effect the whole world. Transferring this information is done through two channels, hardware and software. The hardware is the servant of software, for it is software that is the brain that manipulates the hardware. Professionals in the library and information profession need knowledge in both areas to perform with skill and vision.

In today's information age, are library and information scientists really professionals in the true sense of the word, or are they merely stereotypes, technicians, and secretarial workers? This writer believes, knows, that libraries and information professionals are basic, indispensable agents in the current information society. In general, they are intellectuals and have analytical and critical minds. Mason Gross, the sixteenth president of Rutgers University, speaking of intellectuality, said:

> The intellectual has never been popular because he tends to find serious consideration to alternatives to current mores, whether these alternatives be social, political, religious, or what have you. but the intellectual life is essentially the discovery, contemplation, and evaluation of alternatives, and conformity to any given patterns is its death warrant (1).

There are many intellectuals in the library and information science profession, but many of them are not leaders. More aggressive and productive leaders are needed in the profession. Charles Osborne commented on this in a recent article about academia and spoke of "the call for leaders prepared with deeper and more specialized usable knowledge...this thrust toward training to meet the rapidly changing needs of industry and society" (2). Deeper and more specialized education is an important topic which could lead to dramatic changes in the library profession.

Educational Requirements

Past requirements expected of the library professional, in most institutions, have generally included an M.S. degree from a nationally accredited school. Other requirements in this writer's opinion, should include an internship, as part of the programme, before the degree is granted, and certification and continual recertification after the degrees have been earned, whether these be the Master's or Doctoral degrees. Also, faculty members in library and information science schools should be required, periodically, to take a leave of absence, for perhaps a semester, to work in a library. This might give new perspectives that faculty members would use when they returned to work.

Another suggestion is to approach other professional schools, medical, legal, engineering and others, to try to establish working relationships with them and exchange ideas. This relationship could help to develop more specialization in the library and information science profession with participants working together on collaborative classes and projects. There are many ways to do this today, with the various forms and methods of distance teaching and learning, including classes that may be taken through televised lectures and do not require personal attendance of the traditional classroom class.

Continuing Education

Continuing education and research for all professions are especially important necessities today in this age of rapidly changing times. The traditional structural curriculum, with required class attendance, will be altered and supplemented. Academics and professionals will have access to flexible educational choices for learning and for opportunities to put the knowledge to work.

Research

In the area of research, the library professional and library schools have little evidence of quantity or quality in the field. There are some exceptions to this, but on the whole, the production has been limited. In the early history of the profession, extensive research was not required, and, in the professional schools, faculty members often had such heavy teaching loads that they had little time for research. Also, research was not required for tenure. Today there is urgent need for research and more in-depth research in order to solve many problems and to produce new information. This is essential on local, national, and international levels.

Included in the research process should be studies of all knowledge and information that society considers to be basic such as history, literature, language, science - all information. Perhaps, because the computer has created so much change, more attention should be given to scientific and technical information than in former years. World security and safety may be threatened by the quick transfer, in a matter of seconds, of information through computer channels. Libraries, in their information services, might provide life-saving information.

A recent article by Charles McClure and Ann Bishop addresses "The State of Research in Library/Information Science," and identifies five key issues affecting the status of research: image and importance of research; research competencies; strengthening commitment to research in professional associations; communications between researchers and practitioners; and research on the status of research (3). All of these issues are important and deserve careful attention. In dealing with limited and poor research records of the library and information science profession, this writer suggests, as she has several times in the past, that the whole profession become involved in solving its problems.

Problem Solving Meetings

In trying to solve problems and forge ahead, library and information professionals might hold local, state, regional, national, and international meetings with concentration on the particular issues and with recommendations for change. These meetings might be labeled "retreats," with intense attention focused on the problems. Suggested issues might deal with research and in-depth research, with more specialization in the curricula of library schools, with continuing education, certification and re-certification for library and information professionals, and with financial funding sources for these programmes.

Funding

One of the important topics for consideration is obtaining funding; first, for holding initial exploratory meetings; second, for in-depth research studies; and third, for future follow-up work. Funding for several research centers could be a major project. These might include one or two national centers and one or more international centers. It would be desirable to establish at least one of these as a permanent center for research, for exchange of ideas, and for the continuing improvement of library and information science services.

Possible sources of funding in the United States are government agencies such as the Department of Education, the National Endowment for the Humanities, and the National Science Foundation. Professional library associations may also contribute to studies and activities and private donors often give substantial funds for projects which interest them.

Anne Mathews, Director, Department of Education Programmes in the U.S. Department of Education, listed four priority areas to guide the selection of proposals funded by her department in 1989: "These priorities announced in the July 1988 *Federal Register* are information needs; library funding and economics; access to information; and libraries and education" (4).

Leaders Needed

For real progress in the library and information science profession, dramatic changes are essential. An effort should be made to recruit or develop better people, more intelligent people, more people with leadership ability and more entrepreneurs into the field. The profession needs more people who are prepared to show the same initiative as entrepreneurs in other fields. The profession needs people who are willing to "go out on a limb," who are "far-sighted"; people who are willing to stand alone and to risk failure. Present library and information science professionals have such leaders, and they should serve as mentors to younger professionals.

Summary

The library/information science profession has undergone many changes within recent years and in the opinion of this writer, more changes and progress are mandatory in order to meet the current and future information needs of society. These changes include:

1. substantial change in the education preparation of the professional people, in their continuing education, and in more specialization and sharing with other educational fields;
2. doing more research and more continued "in-depth" research and centers for research;
3. having more meetings (retreats) to plan for change;
4. acquiring substantial funding to finance in-depth research and centers for research; and,
5. working to attract more innovative leaders and entrepreneurs who will work hard for the progress of the profession.

In these processes, it is hoped that the profession will strive to involve all library/information science individuals and groups, ranging from local, to national, to international associations and to all information users who will, in a final analysis, be benefactors, for it is a basic fact that information services are vital to world progress and world peace.

References

1. McCormick, Richard P. and Richard Schlatter. *The Selected Speeches of Mason Welch Gross*. New Brunswick, NJ: University Publications, Rutgers University, 1980. p. 51.
2. Osborn, Charles B. "The Structure of the Scholarly Communication Systems," *College and Research Libraries* 50 (May, 1989). p. 281.
3. McClure, Charles and Ann Bishop. "The State of Research in Library/Information Science," *College and Research Libraries* 50 (March, 1989). p. 127.
4. Mathews, Anne J. "An Overview of Issues, Proposals, and Products in Library/Information Research", *Journal of Education for Library and Information Science* 29: 254 (Spring 1989).

THE CONCEPT OF A UNIVERSITY OF THE WORLD

Joseph Becker
President
Universityof the World
13585 Romany Drive, Pacific Palisades, California, USA

ABSTRACT

Long-distance education is possible today due to advancements and improvements in international telecommunications. This paper sets forth the concept of a "University of the World" in which students use computers, radio, and television to take courses in their chosen discipline and, in return, receive course credit and even academic degrees. The author describes the concept, its educational goals, and objectives, the availability of courseware, the impact of global instructional technology, and questions of cost and technological capability. The organization, governance, funding, and operation of the University of the World is also described, including remarks on the benefits of membership and the functional value as seen from the student's perspective.

I was first introduced to the notion of an international information network at a University of Colorado summer study held in Boulder, Colorado, in 1966 under the auspices of EDUCOM, the Interuniversity Communications Council. One hundred and ninety educators, communication specialists, librarians, information scientists, computer scientists, and other professionals came to Boulder for a month to talk about the national and international opportunities made possible in higher education by emerging new technology. The result was a book called EDUNET, published in the Becker and Hayes Information Science Series by John Wiley & Son, in 1967 (1).

Shortly thereafter, in 1968, I co-authored with Wally Olsen the first chapter on the subject of information networks for the *Annual Review of Information Science and Technology* (2). In it, we reported that the literature was alive with references to budding plans for all types of networks, including international networks.

In 1970 I was Chairman of the U.S. Office of Education Conference on Interlibrary Communications and Information Networks held at Airlie House in Warrenton, Virginia (3). One hundred and fifty people of every professional stripe attended. They concluded the week-long conference by recommending to the U.S. National Commission on Libraries and Information Science that it support and encourage the evolution of networks of every type and coordinate the development of U.S. library and information networks with those of other countries in order to facilitate information transfer on an international scale.

During the intervening 20 years, we have witnessed the development of hundreds of digital data computer networks worldwide and watched their gradual interconnection with television and telecommunications networks. Today, we are on the threshold of a new information era of great social significance. For the very first time, we can think realistically in terms of sharing knowledge resources in all media on an international basis. We are in position to begin balancing information supply and demand among the peoples of the world. We are at the point of advancing teaching and learning by coupling educational systems with interdependent global, multimedia telecommunications systems.

The concept of a University of the World originated with James Grier Miller, a scientist and educator who formerly served on the faculties of Harvard University, the University of Chicago, the University of Michigan, and Johns Hopkins University, and was President of the University of Louisville. Two American Society for Information

Science (ASIS) members, Eugene Garfield and Manfred Kochen, have over the years also presented the idea of a "World Brain" at annual conferences and in the literature.

In 1982 Miller founded an organization with the name "University of the World" and opened offices in La Jolla, California. His purpose was to create an organizational apparatus that would link and connect existing educational institutions in all countries so that the consortium as a whole could serve the educational needs of students in all countries. He felt that the time was ripe to consider educational systems from an international and global perspective; he believed that the need for education transcended national borders, and he held the opinion that the doctrine of *equal access to knowledge* applied to all people.

Members of the University of the World's Board of Trustees include Ambassador Jose Maria Chaves; Ambassador Jean Girard; Lord Walter Perry, Vice-Chancellor of the British Open University; Dennis Jennings, Executive Director of the European Academic and Research Network; Bela Banathy, Associate Director of the Far West laboratory for Educational Research and Development; Kjell Samuelson, Chairman of the Programme of Informatics and Systems Science at Stockholm University; and William C. Hittinger, Executive Vice President for Research and Development at the RCA Corporation.

During the past few years, as the concept took shape, alternative names for the University of the World were suggested. Among them, for example, were "electronic global university," "open electronic university," and "world electronic university." In spite of the fact that the concept implies electronic capabilities, we prefer and have retained the more generic name, *Universityof the World,* because it emphasizes content over process and underscores our belief that *what* is communicated is more important than *how* it is communicated.

Purpose

The overarching goal of the University of the World is to advance free speech and thought and to promote worldwide literacy and scholarship, on the premise that a better-educated world will yield improved international understanding and offer all people a better chance for peace and prosperity.

From the start, the main focus of the University of the World has been on the student. The student is the principal user and the ultimate beneficiary of the services to be offered by the University of the World. To meet individual student requirements, various levels of education would be available, for credit or non-credit.

It is not expected that the University of the World will grant academic credits or degrees of its own. Rather it would make arrangements whereby presently established institutions of higher education, willing to give credit for learning by instructional technologies, could do so internationally. The University of the World would be a facilitating organization and not a campus institution in the classical sense.

Most of the world's population growth in the foreseeable future will occur in the developing countries, many of which cannot now afford adequate primary and secondary education by traditional means. They cannot train a sufficient number of teachers even to minimal standards. Unless mass education by modern technologies is employed, these countries may never be able to take their proper place in the world and provide a reasonable quality of life for their citizens. Therefore, the University of the World has been designed to use electronic telecommunications to disseminate educational resources to students and faculty in all countries, especially the less developed nations. We believe computer-aided instruction systems and video information systems are now sufficiently perfected to enable the multiway sharing and delivery of course material in a broad set of academic fields.

During the past few years, the concept of a university of the world has been discussed with educators, chancellors, and university presidents, as well as government officials at the levels of ambassador, under-secretary, and foreign minister. All have expressed their support and enthusiasm for the concept.

One prime minister put it this way: "Nothing can be more pacific than having students in all countries learn from professors in all lands."

In a letter to the University of the World dated August 8, 1986, Under-Secretary A. Wallis of the U.S. Department of State wrote:

> The United States actively promotes the development of free and open communication links. In the 1985 OECD Declaration of Transborder Data Flows, for example, we pledged to keep the international exchange of information as free as possible, promote access to data, information, and related services, and avoid creation of unjustified barriers. We believe that advanced communication technologies have demonstrated their potential to serve as teaching instruments and to bring about important economic and social benefits.

The Director General of Unesco, Federico Mayor, in a letter to Dr Miller a few months ago, noted that the aims of the University of the World are compatible and consistent with the educational aims of Unesco. He wrote:

> The idea of an international education and research network, putting to use all the benefits of modern communications and information technology, is very much in line with Unesco's own outlook and its search for ways to improve access to knowledge in today's world...[University of the World] goals are indeed in harmony with those of Unesco whose Constitution stipulated *inter-alia* that there should be free and equal opportunity for education for all and a free exchange of ideas and knowledge.

Distance Education (4)

Educators in most countries have long recognized the need to build strong, high-quality national educational infrastructures. Over the years, each country has developed internal programmes and independent means for educating its own people. As a result, our present-day world encompasses numerous disparate educational resources confined mostly to local use. These resources are of mixed quality, of uneven strength, and not always available where the need is the greatest.

The term *"distance education"* has emerged to denote the process by which educational interaction occurs between student and teacher in different geographic locations.

Early examples of distance education include correspondence schools which used the mails for student-teacher communication beginning in the 1930s. Growth in distance education was sustained through the 1970s, and it gradually led to open universities in different parts of the world.

The open university uses the mail, radio, and television for the dissemination of course work. At least 30 other countries have educational television activities today, including the Soviet Union, China, Japan, Canada, and the United States.

At Boston University, for example, a student outside the institution but enrolled in its engineering and computer science television courses has the same option to earn credit and an advanced degree as does the on-site student. Student engineers can thus obtain a master's degree without ever setting foot on the campus. They receive videotapes from the university and send back their homework and examinations for grading. If they have questions, they can speak to their professors by telephone or discuss course content with them through interactive video. Despite the intervention of machines, shared human values, high ethical standards, and high-quality education are not adversely affected by the process.

It is clear that the future of international education lies well beyond the use of the postal service and the telephone. We are currently entering a period in human history when the greater portion of the information generated each day will be electronic. As personal computers, TV monitors, and video information systems spread throughout the world, we can expect even more information to be generated in digital language and to appear in electronic form. This development implies increased reliance on telecommunication systems to carry the traffic.

Today, no well-organized international electronic network exists with clear-cut academic goals for connecting educational and research organizations into a global university network which students of all countries can contact to share human and material resources and learn from professors in all lands. This is the goal which the University of the World hopes to achieve.

The principal ingredients needed to reach this goal are: 1) quality courseware; 2) suitable instructional technology, including an accessible broad-bandwidth international telecommunications network; and, 3) an organization that can plan and govern the operation of the network.

Let us address these issues one at a time.

Courseware

Excellent educational courseware is being developed in different countries. Included among them are many full courses and briefer modules of computer-aided instruction and video programmes such as Carl Sagan's "Cosmos,"Lord Clark's "Civilization," and Bronowski's "Ascent of Man."

Innovations in educational video have also been springing up rapidly. Building on the experience of open universities, many universities and educational consortia have extensive plans for generating computer and video courses in many subject fields and distributing them by satellite. Now that broad-bandwidth satellite communications have become affordable, these plans include provisions for broadcasting this courseware internationally. Among the more prominent organizations concerned with the distribution of courseware by satellite are the following:

- *PACE (A European Programme of Advanced Continuing Education).* PACE is headquartered in Paris. It is sponsored by five major industrial organizations: British Telecom, IBM, Hewlett-Packard, Philips, and Thomson. Its purpose is to offer courses for non-credit and credit toward advanced degrees and for postdoctoral work. PACE aims at establishing an educational network between and among technical employees in industry and the higher education community in Europe. To some extent, PACE is an effort analogous to the National Technological University, which has operated in Oklahoma since 1985.
- *NTU (National Technological University).* NTU was founded to serve the advanced education needs of scientists, engineers, and technical managers. It shares a domestic satellite network with the Association for Media-Based Continuing Education for Engineers (AMCEE). NTU and AMCEE began regular satellite delivery of advanced technical education in August 1985. Today, NTU offers more than 4000 hours of academic credit instruction; AMCEE offers over 1500 hours of non-credit instruction over the same network.
- *FISCIT (The Foundation for International Exchange of Scientific and Cultural Information by Telecommunications).* FISCIT was established a few years ago to plan for new approaches to higher education by long-distance transmission among faculties of universities in Europe and the United States.

The U.S. Department of the Army, which has operated an active, long-standing correspondence programme for training enlisted and officer reserve personnel, is currently investigating the use of satellite information networks and teleconferencing to achieve the same objective.

These pioneering activities clearly demonstrate an ability to produce courseware for distance education purposes in specialized fields. I think they also attest to the fact that a University of the World can be academically and technically viable.

Instructional Technology

As envisaged, the University of the World would be an educational linkage of academic institutions concerned with the storage, retrieval, and exchange of academic information for instruction and research at all educational levels and in all disciplines. It would not rely solely on interactive television and computers, but would incorporate all forms of electronic instructional technologies, such as computers, television, laser disks, and telecommunications, to achieve educational objectives. Nor would it rely solely on specific means of communications. Rather, it would offer all available means for achieving two-way, interactive communication and teleconferencing between and among faculty and students at educational institutions in the network.

Computers

Students and faculty participants in the University of the World would employ computer conferencing and electronic mail techniques to communicate with one another. Two significant networks are already in use to facilitate a variety of university collaborative conversations throughout the world.

The first is called BITNET. It is administered by EDUCOM in the United States. BITNET is a user-oriented international computer network which functions as a store-and-forward electronic mail and message system connecting hundreds of universities in the United States and Asia. In time, it will provide access to many computerized databases, computer-aided instruction, course directories, and other services useful to institutions of higher learning.

Another electronic mail and message system is a large European network known as EARN (European Academic and Research Network). It too is a store-and-forward computer network connecting academic institutions, universities, and research centers in Europe, the Middle East,and very soon, in Africa.

BITNET and EARN are connected to each other and to other networks in other countries such as, for example, NETNORTH in Canada, and JANET in the U.K. Thousands of users in more than 30 countries send thousands of messages each day to one another over BITNET and EARN.

Because BITNET and EARN are comprehensive networks and are up and running, the University of the World intends to use their facilities for both planning and initial operations.

Telecommunications

As the decentralization of small personal computers and TV monitors extends, more and more people will be able to communicate with various forms of information from a distance. When computer and television technology are interconnected in the same telecommunications system, the combination will provide a wholly new educational environment for studentteacher interaction in distance education. The availability of this new telecommunication capability makes the operation of organizations like the University of the World both possible and feasible.

It is clear that the future of international education lies well beyond the use of the postal service and the telephone. Communications are needed to bridge the physical distances between educational institutions and between faculty members and students. Communications are needed to carry the digital and analogue signals bi-directionally over the same lines and to provide channel capacities great enough to handle the enormous volume of international education message traffic implied by an operating University of the World.

Some countries recognized these communication requirements as early as the mid-1960s, and steps were taken then to upgrade terrestrial and satellite telecommunications systems. That is why narrow-bandwidth telephone lines are gradually being replaced by broad-bandwidth coaxial cables, fiber optic pathways, microwave links, and communication satellites. The aim is to install operational, point-to-point, broad-bandwidth telecommunications lines which can move educational information in electronic form back and forth efficiently. Narrow-bandwidth telephone wires, originally designed to carry voice signals, are unable to carry data traffic efficiently from tomorrow's education machines, computers, television, facsimile, compact discs and video discs, and so forth, all require broad-bandwidth channels for their efficient communication.

Once broad-bandwidth channels are in widespread use internationally and nationally, a totally new integrated telecommunication infrastructure will be available for the global exchange of educational information. A broad-bandwidth telecommunications system will make it possible to mix signals bi-directionally and to accommodate different types of machines on the same line at the same time.

The establishment of broad-bandwidth telecommunications systems throughout the world heralds a dramatic change in the international education movement. They will enable educational institutions in every country to make greater use of electronic information created throughout the world by providing the necessary lines, connections, and interconnections for switching, distributing, accessing, and using educational information. It is the availability of broad-bandwidth point-to-point capabilities that will make the operation of organizations like the University of the World possible.

Although the advantages of telecommunications systems have been recognized by educators for many years, operational use of such networks internationally has been hindered in the past by vexing problems of cost and systems planning. Fortunately, the cost of satellite telecommunications has been decreasing steadily over the years, as has the cost of computer power. But a lot of work remains to be done in the area of systems planning.

As educational institutions make greater use of computers, they are sure to discover new opportunities for student learning. With the continued integration of computer and television technologies and the continued application of these technologies to educational needs, worldwide interest in the idea of a University of the World can be expected to grow.

The institutional perspective

Early in 1988, the University of the World sent letters to 77 countries inviting them to join the programme. The letter described the practical benefits of membership. For example, membership in the University of the World implies:

- *Enrollment in BITNET and EARN.* A member country gains access to professors, libraries, and laboratories in various other universities throughout the world.
- *Reduced travel costs.* Fewer foreign trips will be taken by students since course work, in many cases, can be taken without leaving the country. This benefit should also have a salutary effect on the "brain drain" problem endemic to many to the less developed countries.

- *Improved cultural understanding.* An association with the University of the World enables member countries to communicate directly with educators elsewhere in the world who share similar educational values and interests.
- *Cost savings in capital construction.* Once the principle of local self-sufficiency erodes, fewer campus buildings will be needed.
- *A voice in course content and style of presentation.* Membership gives the joining country the right to decide itself on the content of the courses to be taken by its students. Course content, quality, and method of presentation must be culturally acceptable to the member.
- *Conferences, contacts, and communication.* Membership entitles the country to attend the University of the World conferences, participate in international working groups, and receive relevant studies, newsletters, and other pertinent documentation.

The student's perspective

For illustrative purposes, the following is a description of a typical student's learning environment in an international educational network.

The work station would consist of a personal computer, including a laser disc and CD-ROM searchable storage. The computer screen would be split so that it could double as a color TV monitor. The monitor would be capable of receiving TV from the air, videocassettes, and cable.

The student's workstation would be connected to broadbandwidth telecommunication lines that made long-distance connections over satellites. Special low-cost educational tariffs would be paid. Batches of courseware could be transmitted (downloaded) over the network to disc memory at the work station.

The student would be presented with a screen menu that enabled him or her to consult several databases maintained by the central offices of the University of the World. For example, a directory of courseware would be maintained, indexed by subject, by instructor, by university, by date, among others. After the student had selected a course, the system would describe the material and explain the process by which the student can gain access. It would determine whether or not the student was working for academic credit and request registration information. Thereafter, students could be on their own until they completed a course unless they needed to consult with the professor, local experts, or people they could reach on the international network. Reference material would be available on the TV monitor.

The central system would keep track of the students' interaction with the system and maintain a financial accounting of their use and time. Examinations would be given when students were ready, and a computer record kept of the results. Evidence of satisfactory completion of a course would be evaluated by an instructor, who would either authenticate it for academic credit or request the University of the World's central office to issue a certificate of satisfactory completion. A history of the student's course work plus an invoice would be sent to the member country's representative.

The University of the World's main computer would keep a running record of course use, time, cost, and related statistical data that would determine which courses were the most popular, how each country was using the available resources, and whether charges should be increased or decreased.

The human element should not be overlooked. Students at all levels, even when they are using advanced technology, need to be motivated and encouraged by teachers, family, and other students.

Governance

The University of the World recognizes the need for a well-articulated governance policy that will give members a voice in the development of policies, plan, and operations. In this context the term governance means the structure for carrying out consensus actions that do not disturb existing organizational allegiances or interfere with the independence of individual member countries. The governance mechanism for the University of the World must allow for the balanced participation of all those who will be affected.

In addition to an effective governance mechanism, the success of the University of the World will depend on adherence by its members to management principles which motivate people to cooperate and work together. Total unanimity in support of all initiative cannot be expected, but we should try to achieve as much consensus as possible.

The distinction between national independence and functional interdependence must be recognized. University of the World members have independent responsibilities to support their respective countries first and foremost. By joining the University of the World, they are not being asked to accept a conflicting authority structure, but rather to identify and to work on certain functional areas of joint interest whose development will result in major benefits for the group as a whole.

The single most important governance responsibility of all members of the University of the World is setting a positive tone and achieving harmony among members.

Next Objectives

By the end of 1991, the central office of the University of the World expects to:

- Solicit more countries, institutions, and individuals to join the University of the World.
- Assist member countries in establishing a University of the World office of their own.
- Meet with representatives of member countries and assist them in determining the magnitude and substance of their distance educational requirements.
- Meet with BITNET and EARN to establish electronic message capabilities for University of the World members.
- Collect data on courseware, experts in many fields, libraries, and other information of common interest to member countries and enter this data into a central computer database accessible through BITNET or EARN.
- Meet with commercial telecommunications companies to develop the services and probe the costs of operating an international educational network.
- Describe in detail how the University of the World will be organized, governed, operated, and evaluated. Identify problem areas that will require early resolution.
- Develop standards for judging the academic quality of relevant courseware.
- Evolve a financial structure and a method of payment for student use of University of the World services.
- Confer with a select number of universities to determine how students in one country can gain degrees or obtain certificates of completion from member universities in other countries.
- Hold one or more international conferences to bring together government officials, educators, computer scientists, communication specialists, librarians, television engineers, and other representatives of industry. The conferees will discuss issues

relating to finance, organization, cultural considerations, curriculum design, and the application of information technology.

To achieve these objectives the University of the World is actively seeking funds from foreign governments, foundations, corporations, and U.S. grant-giving agencies. Such funds are required for the general administration of a Central Office and for the planned conference.

Conclusion

Benefits to students of an educational programme of the sort we envision include access to excellent teaching from other parts of the world and improved ability to live and work effectively in a world that is characterized by rapidly increasing international interaction and interdependency.

There is now no question that students can learn from television and computers. in fact, evidence is accumulating that learning may be faster and retention better when such media are used for certain subjects in lieu of large classes and lectures. With electronic media students learn at their own pace and can repeat any portion of the instruction that they wish.

Experience has shown that the fear of technological unemployment expressed by some teachers and professors is not well founded. Direct displacement has rarely ever happened in places where educational technology has been employed. Instead, the teachers' job is often redefined, and activities and schedule are sometimes altered, but generally they are left with more time to spend on individualized interaction with students for learning and motivation purposes. Professors also enjoy more time for their own research. As new methods become more widespread, however, it is important that administrators in education consider the economic welfare of long-term, dedicated teaching professionals, so they do not suffer from technological unemployment or underemployment.

Some education administrators are skeptical that technology can produce significant reductions in costs. They have an unfortunate tendency to look upon the costs of courseware, hardware, and communications as being expenses to be added on top of all other operating costs, when in reality, what is needed is a total systems approach before such conclusions can be intelligently drawn.

As an information scientist, I view the concept of a University of the World as a total system in which people and machines work together as a team to improve individual learning and benefit communities throughout the world.

As we speak, people are arriving on our planet faster than we can educate them. Because traditional systems of education were not designed to handle billions of people, new means of mass education must be found to supplement them. The time has come for educators and information scientists alike to explore alternative means. There is no doubt in my mind that some variation of the concept I have described today will come about. Hasten the day!

References

1. Brown, G.W., J.G. Miller, and T. Keenan. 1967. *EDUNET*. New York: John Wiley & Sons.
2. Becker, J., and W. Olsen. 1969. In: *Annual Review of Information Science and Technology*, vol. 3, ed. C. Cuadra. Chicago: Encyclopedia Britannica.
3. In: *Interlibrary Communications and Information Networks*. Proceedings of the Conference. Chicago: American Library Association.

4. *Developing Distance Education.* Report from the 14th World Conference. Oslo: International Council for Distance Education.

Note: This paper is based on remarks made by Joseph Becker at the Fourteenth Conference on Distance Education, August 12, 1988 in Oslo, Norway. Mr Becker acknowledges and appreciates the many contributions to the text made by Dr James G. Miller, Chairman of the Board, University of the World.

This paper was first published in *The Information Society* 6: 83-92, 1989 and has been reproduced by kind permission of © Taylor and Francis, UK.

PART II

CONTINUING PROFESSIONAL ACTIVITIES AND PROGRAMMES WORLDWIDE

CONTINUING PROFESSIONAL EDUCATION ACTIVITIES AND PROGRAMMES WORLDWIDE

AFRICA

Continuing Education for Library and Information Personnel in South Africa

C. M. Vink
Medical University of South Africa, 70 Ingersol Road, Lynnwood Glen, Pretoria 0081, South Africa

The South African Institute for Librarianship and Information Science (SAILIS) is involved in promoting the continuing education of library and information workers in South Africa. Many other organizations are also involved, e.g. employers, departments of librarianship.

Committee for Non-Formal Training

The Committee for Non-Formal Training of SAILIS gives specific attention to continuing education. Its functions are:

- identifying training needs;
- gathering and distributing information about training;
- initiating and encouraging continuing education where existing training does not satisfy needs;
- compiling and distributing standards and guidelines for the organization of courses;
- evaluating requests for financial support and providing funds; and,
- obtaining reports about non-formal training presented with the financial assistance of SAILIS, and assisting with the evaluation where necessary.

The membership includes the Chairman and five members of the Committee who are appointed by the Council of SAILIS and the President, Vice-President and Secretary of SAILIS also serve ex-officio on the Committee. The Committee can present training activities, but usually encourages libraries, departments of librarianship, individual members of SAILIS and also sometimes outside organizations to arrange activities. A network of interested individuals has been established in the regional branches of SAILIS to promote, encourage and assist with non-formal training and continuing education.

The Committee budgets annually for training activities and considers requests for financial support of activities. Organizers are, however, expected to recover costs as far as possible by means of, for example, registration fees.

Attendance certificates for training activities are obtainable from SAILIS and can be issued to participants by organizers of such activities.

An annual award for non-formal training was instituted. SAILIS presents the annual award to a person with significant achievements in this field.

Training Activities

The presentation of courses, symposia, and other training activities is mostly done by individual institutions such as libraries and departments of librarianship. Activities take place regularly on a very wide range of topics of interest to the library and information profession.

SAILIS branches regularly hold meetings, talks and training activities of a general nature, while interest groups within the branches devote attention to more specialized topics, e.g. interest groups in the Northern Transvaal Branch include management, community information, training, research and collection development.

SAILIS publishes a monthly newsletter that includes a calendar of forthcoming conferences, symposia, courses, and other training activities. There are also regular reports on such activities in the newsletter, and during 1989 there was correspondence debating whether there were too many continuing education activities. Correspondents strongly supported continuing education emphasizing that it has a vital role to play in the training of library and information personnel.

The University of South Africa, which is a correspondence university, makes available parts of its existing librarianship courses for non-degree purposes to people who wish to update their knowledge.

The READ organization, which is involved in school libraries in developing communities, runs an extensive training programme for teacher-librarians in these communities.

In 1987 the library of the Medical University of Southern Africa presented a symposium on continuing education that dealt with the needs, aims, programmes, methods, and problems related to continuing education. The proceedings were published.

Use is being made of teleconference equipment, often consisting of an office-type telephone with a loudspeaker, for training purposes. South Africa is a country of large distances and so this technique is particularly useful. It can result in considerable cost savings and enable contact and training where this would otherwise not be possible. Teleconferences have been held locally as well as with overseas speakers.

Occasionally, overseas visitors are brought to South Africa to speak at conferences and symposia and South Africans also regularly attend conferences abroad. This contact is valuable in gaining different perspectives on important issues and also for the exchange of knowledge and expertise.

Several books have been published locally on library topics such as the media center, library orientation, using libraries, and bibliographic control. They enable a large number of people to be reached and are intended primarily to suit local requirements.

Useful addresses for continuing education in South Africa:

Mr. C.M. Vink,
Chairman, SAILIS Committee
for Non-formal Training
Medical University of Southern Africa,
P.O. Medunsa
0204
South Africa

The Head
Department of Library and
Information Science
University of South Africa
P.O. Box 392
Pretoria
0001
South Africa

The Honorary Secretary,
S.A. Institute for Librarianship
and Information Science
P.O. Box 36575
Menlopark
Pretoria
0102, South Africa

READ Educational Trust
P.O. Box 30994
Braamfontein
2017
South Africa

CONTINUING PROFESSIONAL EDUCATION IN ZIMBABWE

S. M. Made
University Librarian
University of Zimbabwe, P.O. Box MP 45,
Mount Pleasant, Harare, Zimbabwe

The *objectives of continuing education* in Zimbabwe are:

- to keep abreast and up to date with new developments;
- to develop and maintain competence;
- to widen experience and practical knowledge in the practice of librarianship; and,
- to promote personal job satisfaction and to enhance existing qualifications.

Activities in Zimbabwe

Activities in Zimbabwe include two activities. The Zimbabwe Library Association has a sub-committee in its Council to organize courses for its practicing librarians who cannot get the chance to attend courses outside the country, and the National Library and Documentation Service arranges two-week in-service training courses for its staff in the various library units of which there are more than one hundred. For such programmes, staff are consulted regarding their requirements and their assessment of their training. This will improve the communication structure within the library. Staff are encouraged to attend the courses and made to understand why it is desirable to do so.

Another continuing education opportunity is offered through the *University of Zimbabwe Continuing Education Programme.*

This is an ongoing exercise within the professional activities of the library staff. The programme has become more specific and affords opportunities especially to the middle and lower levels of professional staff to attend courses, conferences, seminars and workshops to enhance their practice. This programme was further enhanced by the donation from a former University Librarian, Mr D.H. Varley, of a sum to be used for continuing education of staff.

Continuing education in librarianship in Zimbabwe is still largely at three levels:

1. In-service training and part-time courses for para-professionals given in various institutions such as the University Library and the NLDS.
2. Diploma courses for non-graduates which are offered at the Harare Polytechnic over a three-year period. This is intended to produce librarians who will fill posts at middle management level, particularly those in public libraries within urban and rural areas.
3. Short-term courses which are conducted for such categories as school library assistants and library assistants in other small libraries.

ASIA

TWO IN-SERVICE TRAINING COURSES IN PAKISTAN

Syed Jalaluddin Haider
Associate Professor and Chairman
Department of Library and Information Science
University of Karachi, Karachi 32, Pakistan

Two in-service training courses were held in 1989 in Pakistan.

First Course: A short course on the "Use of Microcomputers in Libraries" was conducted from August 19 to 24 at the Computer Laboratory of the Institute of Chemical Engineering and Technology, University of the Panjab. Thirty-five working librarians participated. The topics covered basic microcomputer hardware; fundamentals of microcomputer software; microcomputer use in libraries; cataloguing on microcomputers; handling acquisition with microcomputers; microcomputer-based circulation systems; use of microcomputers in reference services; microcomputer-based serial control.

Second Course: Another course, "Training in Management for Library and Information Scientists", was organized from the November 26 to 30 by the National Institute of Public Administration, Karachi. Eighteen working librarians from Karachi participated in this programme.

CONTINUING PROFESSIONAL EDUCATION IN CHINA

Sun Yun-chou
Department of Library and Information Science
Huadong Teachers University, Shanghai, China

Before beginning, it is necessary to define the different terms related to continuing professional education (CPE). In China, lifelong education is the broad term used for all kinds and models of education and training, including regular and continuing education. The term "professional education" is used for education in library and information science and archival science. Formal education is provided in two to three year programmes in secondary professional schools, or two years in colleges, or four years in universities. Students are admitted through national uniform examinations. Usually they live on campus during their school years. They take courses in politics, philosophy, language, mathematics, history, and literature besides the professional courses.

Continuing education is provided for in-service professional training in colleges and universities. In this article, this term will be used for both secondary level (middle or high school) and higher level in-service education for adults, and excludes regular education. In China, in-service education and adult education are usually taken alternatively, because the in-service staff are always adult. CPE in China includes correspondence education, TV education, night university, vocational university, and various short-term professional courses.

Continuing professional education has been very important in China. This article covers the period from 1949 to 1989.

China is a big country with different library systems. According to official statistics from the Department of Culture, the number of libraries and staff in 1986 is as follows:

	Libraries	**Staff**
Public libraries	2,046	31,000
College and University libraries	1,053	32,700
Academia Sinica libraries	140	3,000
Trade Union libraries	25,000	203,781
Special libraries	4,000	

Of course, so many libraries need professional schools to provide them with enough staff.

What is the situation of regular professional education in China? At the higher level, up till 1978 there were in the whole country only two library schools: the Departments of Library Science of Peking University and of Wuhan University. After 1978, the number of professional schools grew rapidly. By 1989, there were 52 teaching units. Even if each unit graduated 40 students each year, 50 units can still only provide 2,000 professional staff per year.

Since China needs different levels of professional staff, secondary level (middle school) professional education is essential. The first such school was not established until 1978. At present there are only 21 secondary professional schools. The regular professional schools can hardly supply the necessary professional staff for libraries. What is the solution to this problem? I believe the most effective way is to walk on two legs. Let CPE go together with regular professional education.

Many different types of continuing education experiences are provided in China. These are the following:

Correspondence education. This is given at both secondary and higher level. High level correspondence education was established in the 1960s by Peking and Wuhan Universities. After three years of study, students receive a college graduate certificate (equivalent to two years of regular education). Now there are many centers for correspondence education. Each center covers a number of cities. Altogether the centers cover a wide area including the border provinces of the country. Students are limited to their chosen city, and should pass the examination held by the center. The center provides a tutor station in each city, and a textbook. Each semester, teachers visit each tutor station for three or four days to help students review the whole course. The tutor station also has its own teachers to help students with the weekly lessons.

Secondary level correspondence education was established in the 1980s with three important centers all run by the provincial libraries in Jilin, Szechuan, and Kwansi provinces. The largest is Jilin Provincial Library Center which is reported to have trained 1,000 students. By 1989, the graduates of these centers numbered more than 5,300.

*TV university.*The introduction of new technology revolutionized the model of education and brought about the development of TV education. The TV professional university started in 1985 with enrolment of 20,000, the largest enrolment in professional education. Students were chosen by the national uniform examination and must pass the final national uniform examination before graduation. Numerous TV work stations were established throughout the country. Special textbooks were compiled, and educational programs were developed. Some people criticize the TV work stations, but the TV University maintained strict standards. In 1988, only half of the students, 10,000 were qualified for graduation.

Night universities. Night universities are run by regular professional schools. The students are mostly adult in-service staff. The textbooks and teachers are the same as those for regular college students, but the students only use the campus at night. Night universities are operated only by a few professional schools.

*Vocational universities.*These universities are usually organized for a purpose different from that of regular professional schools. Some are organized only for the staff of a certain library. Some receive credit for the study program.

Special short courses. Courses are varied in level, scope, and duration. Students may be chief librarians, heads of departments or divisions, or ordinary staff. The duration of courses may be one year, half a year, three months, or even one month. The scope of the course may cover the whole field of library and information science or be limited to one special subject such as classification, Western reference books, etc. The Working Committee of China Higher Educational Institutions Libraries of the Ministry of Education has since 1983 organized two types of courses: one for directors of college and university libraries (half a year) and one for non-professional college graduates newly employed in college libraries (one year).

The Administrative Bureau of Libraries of the Ministry of Culture twice organized training courses for provincial library directors and many other courses for staff numbering altogether 545. Some provinces have organized such training courses for library assistants. The 6th Subcommittee of the Committee of Standard Bibliographic Description has organized courses on AACR2.

Libraries, especially public libraries have a role in continuing professional education. Librarians in public libraries do much of the work of organizing CPE. For example, the TV University is run by a vocational university of the National Peking Library. The three important secondary correspondence education centers are all run by provincial libraries.

Now all public libraries have a Research and Training Department. Provincial libraries have responsibility for training staff for county libraries. City libraries have responsibility for training staff for factory libraries, school libraries, and all the union libraries. It is difficult at present for small libraries to get graduates from professional schools. Through short courses organized by the provincial and city libraries, however, the staff of these small libraries can provide themselves with professional knowledge and techniques. According to the statistics of the Administrative Bureau of Libraries of the Department of Culture, up to 1988 the Provincial Bureaus of Culture and the provincial libraries had organized 573 short courses and trained more than 30,000 staff, and the city libraries organized an even greater number.

Two official conferences on CPE have been held because the central government of the People's Republic of China is greatly concerned about CPE. The first, by the Working Committee of China High Educational Institutions Libraries, was held in July 1984 in Nanking. Fifteen departments attended and reviewed special courses for non-professional college graduates newly employed in college libraries.

A Public Library In-Service Education Working Conference for the whole country was held in June 1988 in Bonfu, Anhui Province by the Ministry of Culture. Before the 1988 Conference, a document was issued by the State council PRC on "The reform and development of adult education." It mentions that "adult education is an important part of China's education." "Adult education should meet the special needs of adults: self-study reading habits; emphasis on practice; specially prepared textbooks; flexible study time." The Bonfu Conference emphasized professional training for staff and the importance of professional training before taking jobs. Several kinds of in-service training were discussed, with emphasis on specialized short courses. Encouraged by the government, CPE in China will grow in importance and make great contributions to the country's professional education.

Bibliography

1. Huang, Zongzhong. "On development and reformation of library science education in our country;" In: *Selected Teaching Reference on Information and Library Science*. Shanghai, Shanghai University: Information Management Department, 1984.
2. China Higher Educational Institutions Libraries. Working Committee. *Proceedings*. Nanking, 1984.
3. People's Republic of China. Ministry of Culture. Public Library In-Service Education Working Conference. *Proceedings*.Bonful, 1988.

CONTINUING LIBRARY EDUCATION IN TAIWAN, CHINA

James S. C. Hu
Professor and Dean
Department of Library Science,
National Taiwan University, Taipei, China

Professional library education in Republican China was initiated in 1920 when Miss Mary Elizabeth Wood founded a library school at Boone University, an American Episcopalian institution located in Wuchang of Hupeh Province in Central China.

The first education programme for training professional librarians in the Republic of China on Taiwan was started in 1955 as a Division of the Department of Social Education at Taiwan Provincial Normal College (now National Taiwan Normal University). In 1961, the first department of library science was established at the College of Liberal Arts, National Taiwan University. As of 1989, there are four universities and one college that have undergraduate programmes in library and information science, located in National Taiwan University, National Taiwan Normal University, Fu-Jen Catholic University, Tamkang University, and the World College of Journalism. Of the five, only National Taiwan University presently offers a graduate programme leading to the Master's and Ph.D degrees.

Continuing library education in Taiwan has been provided by various means and channels including workshops, seminars, institutes, etc. conducted by the Library Association of China, library schools, and major public libraries. Following are brief descriptions of each of them.

Continuing Professional Education organized by the Library Association of China (LAC). The LAC plays a leading role in CPE. Since 1956, it has continuously organized summer workshops for professional and non-professional librarians. These workshops were usually divided into four groups with participation by library workers from public, school, special, and college and university libraries. Most of the workshops were held in July and August, ranging from four to thirteen weeks, with six weeks being the average. Courses offered cover a wide range of subjects as demonstrated in Table 1.

It is generally agreed that the LAC's workshops have contributed a great deal to the promotion of librarianship in Taiwan. According to a survey in 1986, a total of 3,295 library workers had taken advantage of the LAC's efforts in CPE during the period 1956-1985 (1).

Continuing Professional Education offered by Library Schools. Continuing Professional Education in this category varies in format and duration in Taiwan. The Department of Library Science at Fu-Jen Catholic University has occasionally provided short summer workshops for its alumni in recent years.

The Department of Educational Media and Library Sciences at Tamkang University constantly held summer seminars and workshops for its alumni and library staffs of its parent institution. The Library Division of the Department of Social Education of the National Taiwan Normal University usually conducted longer summer workshops for school librarians ranging from six weeks to two months.

The Department of Library Science at National Taiwan University has, since 1979, in cooperation with the Library Association of China, offered a Library Automation Seminar each summer for professional librarians mainly from special and college and university libraries. Participants of the two-week seminar have increased steadily, from forty in 1979 to over 120 in 1988 and 1989. Courses offered cover various aspects of library automation as well as information science. Some of these are listed below:

Application of Microcomputers in Libraries
Artificial Intelligence
Authority Files
Automation of Acquisitions
Automation of Cataloguing
Automation of Serials Management
Chinese MARC
Construction of Thesauri
Indexing and Abstracting
Information Technology
Library Networking
Online Searching
Telecommunications

As Dean of the NTU's library school and director of the summer institute of library automation, it is my pleasure to report that the Continuing Professional Education courses offered by library schools were satisfactory.

Continuing Library Education Courses Offered in Workshops Sponsored by the Library Association of China 1956-1985

Course Title	Frequency
Acquisitions Techniques	21
A-V Materials and Production	6
Cataloguing and Classification	33
Children's and Young Adults' Literature	16
Chinese MARC	4
Information Science and MARC	4
Information Systems and Services	1
Introduction to Information Science	2
Introduction to Library Science	20
Library Administration	4
Library Automation	5
Library Development and Future Trends	10
Library Extension Services	2
Library Practical Works	12
Library Programmes and Services to Children	10
Non-Book Materials	15
Public Libraries	9
Reference Services	25
School Libraries	10

Table 1. Continuing Library Education Courses Offered in Workshops Sponsored by the Library Association of China 1956-1985 (2).

Continuing Professional Education Sponsored by Taiwan Provincial Taichung Library. The Taiwan Provincial Taichung Library is the most important public library in the province of Taiwan. It has been legally charged by the Department of Education of the Provincial Government to provide short training classes and Continuing Professional Education workshops for school and small public libraries in the province. Since professional librarians are still lacking in these libraries, their courses are basically general and introductory in nature. In fact, most of the participants are library clerks rather than professionals. Therefore, continuing library education at this level can hardly be regarded as Continuing Professional Education.

To sum up, Continuing Professional Education programmes in Taiwan are primarily provided in the summer and some of the activities are actually training classes for non-professional library workers. Although there is a lot that needs to be done in this type of education, it has been recognized that the past accomplishments have had a fair share in the contribution to the advancement of librarianship in this area.

References

1. Li-hung Huang. "A Study of Continuing Education for University Librarians in the Republic of China." Master's Thesis, National Taiwan University, 1986, p. 37.
2. Shih-hsion Huang. "Continuing Education and Staff Development for Librarians in the Republic of China." In: James S.C. Hu, (ed.) *Library and Information Science Education: An International Symposium.* Taipei, National Taiwan University. Department and Graduate Institute of Library Science, distributed by Scarecrow Press, 1986. p. 213.

CONTINUING PROFESSIONAL EDUCATION IN THAILAND IN 1989

Sa-angsri Phornsuwan
Assistant Professor
Rankhamhaeng University
373 Thonglor 19, Sukumvit 55, Bangkok 10110, Thailand

The Public Library Group of the Thai Library Association conducted a one-day workshop on September 13, 1989 at the Non-Formal Education Department, Ministry of Education. "Services and Activities in Public Libraries" was the topic of this workshop. There were sixty participants from the public libraries.

Thailand is a host country for the programme of exchange librarians from the ASIAN countries: Thailand, Indonesia, Malaysia, the Philippines, and Brunei Barussalam. The National Library of Thailand with the cooperation of the Thai Library Association, academic libraries and schools of library and information science were the host organizers of a programme called "Education and Training for Library and Information Personnel", held from November 19 to 30, 1989.

The Thai Library Association organized its annual conference at the Bangkok Palace Hotel, December 11-14, 1989, on the theme "A National Information System for the Development of Thailand".

CARIBBEAN AREA

CONTINUING PROFESSIONAL EDUCATION FOR LIBRARIANS IN THE ENGLISH-SPEAKING CARIBBEAN, WITH SPECIAL REFERENCE TO TRINIDAD AND TOBAGO

Esahack Mohammed
School Libraries Division
Central Library of Trinidad and Tobago (Tobago)
21 Third Street, Barataria, Trinidad and Tobago

Continuing professional education (CPE) is of critical importance to library and information professionals in the Caribbean. The profession in the region is by no means uniform: it is characterized by division into four main language groups: English, French, Dutch, and Spanish-speaking, and by wide geographic dispersal, practicing as it does in islands spread over a sea area one third the size of continental United States. On the one hand countries with sizeable numbers of professionals can be found, such as Puerto Rico, Jamaica, and Trinidad and Tobago, while on the other there are those where the numbers are quite small. As can be expected, each language group has its own method of initial training and criteria for entry into professional ranks, but even in any one group, and certainly in the English-speaking one, there are differences.

An Overview of Basic Professional Education in the English-speaking Countries

The entire English-speaking area has a tradition of formal professional education derived from the British system which was implanted in the region some four decades ago with the establishment of a School of Librarianship in Port of Spain, Trinidad. The School served to establish a cadre of pioneering professionals for the region as a whole (1). By the early 1960s, the School was phased out and aspiring librarians had to enter library schools in the United Kingdom. Within the last two decades, the most significant development impacting on professional education has been the establishment in 1972 of a full-fledged library school as part of the University of the West Indies (UWI), based at the Mona campus in Jamaica. This school offers both an undergraduate degree and a post-graduate qualification, formerly a diploma, but since 1989 a master's degree in library studies. In the almost twenty years of its existence, the School has trained a very large number of librarians and information personnel now in the English-speaking area. A survey undertaken for this paper in Trinidad and Tobago revealed that 49% of 130 (approximately 65) librarians there are graduates of the school.

Further Education

Only a small percentage of librarians have higher level library qualifications; these have received advanced degrees at either U.K. or U.S. library schools. Factors inhibiting the more widespread pursuit of higher qualifications relate to the limited facilities at the UWI Library School and the highly increased cost of study abroad due to the fall in the value of local currencies vis-a-vis those of metropolitan countries.

Factors Affecting Continuing Education in The Region

Several factors are pertinent to Continuing Education in the region. Some of these relate to:

1. the diversity of working environments, which vary from well-organized centralized systems to school and small specialized libraries operated by one person;
2. the variation in staff numbers among the various islands;
3. the isolation of some staff in the smaller islands and smaller libraries in the larger islands;
4. the absence in some islands of large library systems whose practice can serve as a model for other libraries; and,
5. the youthfulness of the profession.

Among these, the chief determinant of the need for CE must be the youth of the present body of practicing professionals. A random sampling of 10% of the membership of the Library Association of Trinidad and Tobago reveals the age structure of the professionals in that country as follows:

Ages	**% of Membership**
20-29	20%
30-39	55%
40-49	15%
50-59	10%

The implications of this kind of age distribution are of tremendous relevance to the question of the kinds of CE activities to be organized because of the fact that some 75% of the personnel now in the field will be practicing well into the 21st century and some into the second decade of that century! The challenges facing the profession in this region are therefore much wider than keeping abreast of new knowledge and techniques and the maintenance of professional standards. Rather it relates to the harnessing of CE for the development of leadership roles and the enabling of our professionals to become the shapers of an information future, having regard to the developing nature of our societies.

The social context in which our library and information personnel must operate include countries where basic library services have to be brought to citizens on the one hand, while on the other, state-of-the-art technological advances are already in place in some others. Additionally, both ends of the spectrum may be represented within some islands, where library development has proceeded very unevenly over the past few years. The need to maintain parity with international developments in order to have access to internationally available information by technological means is evident if the societies are to be supplied with adequate information to serve development needs. In these contexts, our librarians will be called upon to make critical decisions about the nature of our library services and the use of limited financial and manpower resources.

Providers of Continuing Education

The organizing of continuing professional education in the larger islands like Jamaica and Trinidad and Tobago, where there are upwards of 200 professionals, is a practicable proposition, but this is not so for the smaller islands where numbers are few. Accordingly, CE is quite varied from island to island, but there are common linkages. The providers of CPE fall into four main categories, library associations in the individual islands, regional library associations, the Library School, UWI, and employers.

Library Associations

There are library associations in most of the countries of the Caribbean, and these have been in the forefront of the thrust for CE. In Jamaica and Trinidad and Tobago, where library associations have been in existence for some three decades, and in some other countries like Barbados and Antigua and Barbuda where they were formed more recently, a large proportion of the associations' efforts have been directed towards organizing workshops and seminars either independently or jointly with other agencies. The thrust of these activities has been to update practitioners and help them to keep abreast of new developments. The short course on "The Micro-Computer for Information Work" organized by the Jamaica Library Association in 1985 (2) and a two-week course entitled "New Trends in Information Retrieval", organized by the Department of Library Studies, UWI, and hosted by the Library Association of Trinidad and Tobago (3), are examples of the efforts of library associations to promote awareness of change and the introduction of new ideas and techniques to professionals.

Regional Associations

The most outstanding CE activity for the entire Caribbean region, that is, the English, Dutch, French and Spanish-speaking countries, is the annual conference of the Association of Caribbean University Research and Institutional Libraries (ACURIL) which has been held annually since the formation of the association in 1969. These conferences extend over a one-week period, with pre- and post-conference workshops and seminars and are held in a different country each year. The themes of these conferences are always oriented to the concerns of the region as a whole and are capable of influencing the direction of librarianship and information in the area across language and administrative barriers. The intellectual stimulus of these ACURIL conferences continues beyond the week of activities and influences a much larger audience than conference attendees through the publication of annual conference papers. Countries with large bodies of professionals, such as Jamaica and Trinidad and Tobago, are represented by 8-10% of their staff, while those with smaller numbers are represented by as high as 50% of the professional body. Additionally, the country hosting the conference benefits not only by even higher participation, but by the experience of planning, budgeting, managing, and publicizing the conferences, thus enhancing their competence in these areas of professional practice.

The Department of Library Studies, UWI

Professor Daphne Douglas, Head of the Library School, is of the view that "the library school in the developing country or region must...accept responsibility for continuing education activity and offer a wide range of topics" (4). The perspective of the School is as broad as the English-speaking Caribbean and, therefore, it supports activities designed to have maximum benefit and multiplier effect within the region. Thus, in addition to such facilities as the auditing of courses by individuals, the resources of the School are used to support the work of professional associations in the area of training seminars, workshops, and short courses, usually with several countries participating, and with financial and other inputs from international funding agencies.

It is anticipated that as the resources of the school expand, especially in the area of manpower, it will be in a position to assume a more prominent role in the provision of CPE for the region. This new role will enable the School to develop and maintain a lifelong professional relationship with its graduates. Offerings need not be delivered only at the school's campus in Jamaica but should be available at other campus sites, especially during the long summer vacations. A very pertinent factor in considering an expanded CPE role for the School is the fact that it is conceivable that in the not too distant future the School may achieve an output of graduates beyond the capacity of the region to absorb in the short run. Should this appear likely to occur, the onus on the School to deliver more CPE cannot be overstated.

Employers

Librarians employed in public, school and a large number of special libraries comprise an integral part of the government's public service. They are thereby enabled to participate in a wide range of general training activities, organized by centralized training agencies of the government. Thus librarians participate, along with other professionals, in such courses as Personnel Management, Performance Appraisal, and Management in the Public Service. The grouping together of several professional participants, representing different areas of the public service, serves to enhance the ability of librarians to deal with colleagues in other areas and likewise fosters an appreciation of the work of librarians among those other areas and professions represented.

One of the most fruitful innovations in CE was made some eight years ago by the Central Library, the public library service employing about 15% of professionals in the country. This involves the convening of regular bi-monthly meetings, devoted wholly to activities designed to promote professional development. Meetings, which are chaired by younger members of staff, are planned on a calendar-year basis, with participants deciding on the topics to be explored. These may be on library or other job-related activities. Thus there have been lectures on stress management and community outreach library services, and reports on significant literature, courses, and conferences attended. There is a high level of interest and participation, which is attributable to the fact that members themselves choose the topics to be addressed. Proceedings of these meetings are edited and are intended to be published for wider circulation.

Another series of meetings of a similar nature is organized by the Schools Library Service, a Division of the Central Library. These meetings are held once each school term, specifically for school librarians working in secondary schools throughout the country, who comprise about 20% of the overall librarian population of Trinidad and Tobago. These meetings are of great importance to the school librarians, who work largely isolated in the schools, for keeping in contact, discussing problems, describing innovative activities, and generally renewing and refreshing their ideas and experiences. Outside speakers and presenters on library or related topics are a regular feature of these meetings.

The Road Ahead

It is evident that while the region is fragmented geographically, there is a great deal of professional unity and harmony, due largely to the existence of a unified professional education system for all of the English-speaking countries. In this context the need for one organization able to provide for the major CPE needs of the entire region will be a distinct advantage.

The one existing agency which can fulfil this need is the Department of Library Studies, UWI, provided that its resources are so structured and deployed as to enable it to assume this role more fully than at present, alongside its existing responsibility for basic education. Appropriate programmes can be devised with input from professional associations.

An important matter to be faced is the question of costs. The existing and projected financial climate would seem to indicate that, more and more, individual librarians will be called upon to shoulder greater financial responsibility for their CE activities. For this to be worth while, professionals would have to be able to receive recognition of their participation, perhaps on a model similar to that of the Continuing Education Units offered by some institutions in the U.S.A.,or on the model of the International Graduate Summer School (IGSS), operated by the College of Librarianship Wales and the School of Library and Information Science, Pittsburgh, where courses are taken for credit. The question of rewards in terms of career advancement will also have to be addressed.

Additionally, having regard to the scattered nature of the islands, the consequent high costs of travel within the Caribbean, and the isolated conditions under which some librarians practice, the situation is ideal for experimentation with the application of Distance Learning to CE efforts in the field. Distance Learning has been in the Caribbean for some time now in experimental form in the UWI Distance Teaching Experiment (UWIDITE), which links the three campuses and the three Extra Mural Centers. The Librarian at the Mona Campus, Albertina Jefferson, says that the network is being used for "interactive distance education university courses and courses leading to university qualifications" and is of the view that "many different forms of CE for professionals are adaptable to dissemination via a telecommunications system" (5). She suggests that programmes could be aimed at upgrading competence in particular subject areas and providing "refresher courses" for professionals.

There thus exists the possibility of taking CE into the future through a combination of traditional means and distance learning via satellite, which will enable it to reach a much wider professional audience than at present. There can, for example, be combinations of short courses and seminars via satellite with attendance a longer summer sessions for CE Units or credits. The peculiar position of these islands calls for innovative approaches by all the present providers of CE, so that a unified approach in the interests of the profession as a whole can be devised. This is imperative because it is CE which will determine to a very large extent the library and information future of the Caribbean.

References

1. Mohammed, Esahack. *The Eastern Caribbean Regional Library School of Librarianship: an Experiment in Library Education*, 1984. Unpublished.
2. Aarons, John A. "The Role of the Jamaica Library Association in Continuing Education for Librarians." ACURIL. XVII Proceedings: *Continuing Education for Librarians in the Caribbean,* USVI, 1986, pp. 95-97.
3. Mohammed, Esahack. "The Role of the Library Association of Trinidad and Tobago in Continuing Education for Librarians." ACURIL, XVII Proceedings: *Continuing Education for Librarians in the Caribbean,* USVI, 1986, pp. 98-100.
4. Douglas, Daphne. "The Role of the Library School in Continuing Education with special reference to developing countries." ACURIL, XVII Proceedings: *Continuing Education for Librarians in the Caribbean,* USVI, 1986, pp. 47-50.
5. Jefferson, Albertina. "Distance Learning as a Means of Continuing Education in the English-speaking Caribbean." ACURIL, XVII Proceedings: *Continuing Education for Librarians in the Caribbean.* USVI, 1986, pp. 105-111.

EUROPE

CONTINUING PROFESSIONAL EDUCATION FOR LIBRARIANS IN THE FEDERAL REPUBLIC OF GERMANY

Jurgen Scheele
Stuettgerhofweg 40, Koln 40 5000, Germany

Librarian in-service education in the Federal Republic of Germany is offered by a multitude of institutions: most commonly by the professional schools and the professional associations, but also by private commercial agencies (1). In recent years, the Fachhochschule für Bibliotheks- und Dokumentationswesen in Cologne offered the most courses in continuing professional education for librarians (2). As of 1990, however, staffing difficulties have forced this institution to drastically reduce its programme, despite a demonstrated need for refresher and in-service education (3).nI recent years, common themes were: introduction to data processing, word processing, database systems, online searching, and other new technologies, as well as cataloguing rules (RAK, RSWK), public relations and, especially for librarians in public libraries, collection building in different subjects and in fiction. Courses in interlibrary loan and bookbinding were offered at regular intervals.

Although, the courses in the past were largely free of charge, except in private institutions, there is a trend toward covering expenses by charging fees.

The profession has repeatedly demanded a central clearing agency for the coordination and planning of continuing professional library education, but no such agency has been established as yet. The Fortbildungskommission des Deutschen Bibliotheksinstituts, a voluntary organization, serves as the only infra-structural agency. This commission is made up of representatives from all library fields and from all regions of the Federal Republic of Germany. The members of the commission observe the continuing education activities in their own regions and exchange information and experiences. They are the regional representatives for all matters concerning continuing professional library education, and help, as far as possible, to find speakers. The Fortbindungskommission develops guidelines for carrying out in-service education (4). It does its utmost to supplement the present fragmentary offerings in continuing education for librarians by ensuring that various topics are covered and that courses are given in various regions (5).

References

1. Fortbildingsadressbuch: Bibliothek Information Dokumentation/hrsg von der Fortbildungskommission des Deutschen Bibliotheksinstituts - 2. Uberarb. Aufl. Berlin 1989 - (Dbi-Materialen: 46). The most important agencies present themselves in Buch und Bibliothek 38 (1986), pp. 224-250.
2. Fortbildungsprogramm/Fachhochschule für Bibliotheks- und Dokumentationswesen in Koln - Koln 1984ff.Jung, Rudolf: Weiterbildung als Aufgabe der Fachhochschulen und moglichkeiten der Realisierung in der Fachhochschule für Bibliotheks- und Dokumentationswesen in Koln - in: Verband Mitteilungsblatt N.F. 33 (1983), pp. 105-114.

Scheele, Jurgen: Fortbildung in Nordrhein-Westfalen. - In: Buch und Bibliothek 38 (1986), pp. 245-250.
Marloth, Heinz: Die Welterbildungsveranstaltungen der Fachhochschule fur Bibliotheks- und Dokumentationswesen in Koln im ersten Halbjahr 1987: Analyse, Bewertung und Ausblick in die Zukunft. - In: Verband der Bibliotheken des Landes Nordrhein-Westfalen. Mitteilungsblatt N.F. 37 (1987), pp. 339-348.

3. Graumann, Sabine: Jurgen Scheele: Fortbildungsbedarf im Informationswesen - Ergebnisse einer Umgrage 1984 - In: Bibliotheksdienst 19 (1985), S. 4-38.
4. - Fortbildungsadressbuch (s. ref. Nr. 1).
   - Hinweise zur Vorbereitung und Durchfuhrung von Fortbildungsveranstaltungen/- hrsg. von der Fortbildungs-kommissiondes Deutschen Bibliotheksinstituts - Berlin 1987.
   - Rechliche Regelungen zur bibliothekarischen Fortbildung/hrsg. von der Fortbildungskommission des Deutschen Bibliotheksinstituts - Berlin 1990.
   - Methodisch-didaktische Grundlagen zur Fortbildung/hrsg. von der Fortbildungskommission des Deutschen Bibliotheksinstituts - Berlin 1991.
5. Fortbildungsplan: Entwurf der Grundsatzplanung zur Fortbildung der Bibliothekare Berlin 1980 (Dbi-Materialien: 3).

REPORT ON THE SYSTEM OF CONTINUING EDUCATION AND EXTENSION TRAINING OF LIBRARIANS IN HUNGARY

Beata Bobok
1025 Csalan Street 49/b
Budapest, Hungary

Education and extension training are constantly developing according to the needs of society. Nowadays everything is changing in our country so we expect a series of reforms in library education in the near future. The present situation is as follows.

The administration of library education is the responsibility of the Ministry of Education and Culture. The Ministry has set up a Standing Committee for Library Education and Extension Training. The members of the Committee are: the Head of the Department of Library Science at the Eötvös Lorand University, Budapest (Chair), the Head of the Department of Methodology, Research and Extension Training of the Center for Library Science and Methodology of the National Széchényi Library (Secretary), Heads of Departments of Librarianship of teacher training colleges, and distinguished representatives of the library profession. Their task is to supervise, help, develop, and coordinate the education, including continuing education, of librarians.

The system of library education and training is spread over three levels, in harmony with IFLA standards:

- University level (general library education and training of information specialists);
- College level (upper and lower grade primary teacher training colleges); and,
- Training of library assistants.

University level education and training takes place at the Department of Library Science, Eötvös Lorand University, Faculty of Arts (1052 Budapest, V., Pesti Barnabás u.1.) and the Department of Kossuth Lajos University, Faculty of Science, (4010 Debrecen.)

Both of these universities offer postgraduate part-time courses. The postgraduate branch of the universities is open to those who have graduated in subject fields other than librarianship, or graduated in librarianship at a college. A person with these qualifications may enter postgraduate courses if they are employed as librarians, information officers, or information specialists, and if their employers support their further training. After three academic years, students receive a degree at the same level as the full time university graduates.

The training of information specialists is provided in another postgraduate course at the Eötvös Lorand University. Studies take two academic years on a part-time basis for applicants with a university qualification, a few years of experience in librarianship and information work, and knowledge of foreign languages. At the completion of their studies they receive a diploma.

The third form of continuing education at the Eötvös Lorand University is a stop-gap, one-year training for school librarians who have already graduated. Graduates receive a certificate on completion of their studies

By continuing education and extension training, we mean both higher level (postgraduate) education and specialized training and refresher courses which do not lead to a new qualification or a degree. Extension training may be either stop-gap training which may be basic training, or supplementary training if basic training has not been sufficiently comprehensive in particular aspects.

Besides the universities, there are institutions which take part in continuing education and extension training such as the National Szechenyi Library - Center for Library Science and Methodology (1827 Budapest, Budavari palota F Epulet). The

Center is the main organizing body for centrally managed extension training courses and also plays a coordinating role in extension training. While its activities are concerned primarily with the extension training of public librarians, its training activity extends to librarians from other types of libraries as well. They organize courses as described below.

Special professional training courses provide high level theoretical and practical knowledge to homogeneous groups of librarians employed in similar jobs. A few examples of the subjects covered at the Center are: children's librarianship, collection development, library pedagogy, classification and coordinate indexing, structure and construction of multilingual thesauri, etc.

Refresher courses offered by the Center include handling of non-book materials, library methodology, etc.

Courses are organized to instruct librarians in the application of new regulations and new library standards. These are called orientation courses and include a course in ISBD-based bibliographic description, and a course held on the occasion of the introduction of a new financing and budgeting system for managers.

The Ministry of Labour and the Ministry of Education and Culture issued a joint decree on the extension training of managers in the field of culture. In response, the Center for Library Science and Methodology periodically organizes management courses for the directors of various types of libraries. They receive up-to-date information on management and library policy. These take one year, and participants receive a certificate of attendance.

Departments of librarianship in teacher training colleges also offer postgraduate training courses. At present they are run by the Bessenyei Gyorgy Teacher Training College (4401 Nyiregyhaza,. Sosto u.31/b), and Berzsenyi Daniel Teacher Training College Szombathely 9701, (Szabadsag ter 4.)

Centers of library networks are also involved in extension training. Their present activity is confined to short conferences of one or two days for the dissemination of new information and new approaches. Some of these centers organize short courses on specific subjects such as on-line user-education given at the National Technical Information Center and Library (1088 Budapest, Reviczky u. 6) or "How to use PRECIS" given at the National Library and Museum of Education (1055 Budapest, Honvéd u. 19)

Bibliography

1. Edit Csapo. *The Education and Extension Training of Librarians in Hungary.* National Szechenyi Library Centre for Library Science and Methodology, 1985, p. 54.

CONTINUING EDUCATION IN ITALY

Alberto Petrucciani
Via A. Carrara 77/6
Genova I-16147, Italy

In Italy, Continuing Professional Education (CPE) programmes are offered by the library association (Associazione Italiana Biblioteche), central, regional, and local government agencies and other organizations, including commercial enterprises.

Degrees and diplomas in library science are offered by only a few universities and these qualifications are not generally required by employers appointing librarians or assistant librarians. Postgraduate two-year programmes are offered by the "Scuola Speciale per Archivisti e Bibliotecari" at the University of Rome and at some "Scuole di Specializzazione" or "di Perfezionamento" at the Universities of Milan, Padua, and Naples, but they are frequently attended by practicing librarians, on an entirely voluntary basis, who wish to gain a better knowledge of library work.

The Associazione Italiana Biblioteche (AIB), as its statute reads, "promotes activities which guarantee professional training and continuing education" of library personnel, by means of conferences, workshops, programmes and courses, and study travel (e.g. to the USSR in 1990).

At the national level, the AIB organizes, for the most part, brief seminars, workshops, etc., on special subjects, sometimes with foreign experts. At the local level, besides special courses, longer basic training programmes may also be offered by the regional sections of the Association, especially if they are not undertaken by the regional or local government agencies. Moreover, the Sections are often involved as organizers or as consultants about contents, schedules, teachers, etc. for programmes initiated and/or sponsored by government bodies and other organizations such as regions, districts, municipalities, public library systems, etc.

The need for continuing education is recognized in laws and regulations relating to the civil service including libraries, so general programmes varying in length from some months to one academic year are often offered. Special courses and workshops may also be offered by organizations in the public sector connected with libraries. At the central government level, some courses are offered by the Ministero per i Beni Culturali e Ambientali, which is in charge of national and other general research libraries, and by the national school for higher civil servants (Scuola Superiore della Pubblica Amministrazaione). At the local level, CPE programmes may be offered by organizations having administrative and planning responsibilities for libraries: regions and districts for public libraries, universities for academic libraries, and the IRRSAEs for teacher-librarians in charge of school libraries, etc.

Sometimes, continuing education programmes and courses are initiated also by other organizations: learned institutions and foundations especially in the field of rare book librarianship and preservation, or associations, chambers of commerce, trade unions, etc. Specialized intensive courses lasting two to six days are also offered by some private companies involved in library science publishing (e.g. the Editrice Bibliografica in Milan) or in consultancy and brokerage (e.g. IFNIA of Florence).

In general, the number of CPE programmes and activities is large, albeit unevenly spread, especially in Southern regions. They are in great demand and well supported by librarians. Continuing education is very necessary because of the lack of formal professional training of people entering the profession, the rapidly changing context of library work, and, in most libraries, the lack of senior or specialist supervision.

However, because of the uneven background of participants and the lack of time for practical work, continuing education activities are not always completely successful.

For special courses and workshops, the subjects most often offered are: library management; library automation and information technologies; descriptive cataloguing, indexing and classification; online information services. The teaching methodology generally involves case work, group discussion, and lectures, with the usual teaching aids such as flowcharts, transparencies, etc. Teachers are usually experienced librarians or sometimes library science professors and experts from other fields such as computer science or management.

References

1. Antonella Aquilina D'Amore (ed.) *Seminario sulla didattica della biblioteconomia*, Roma, (15-19 giugno 1981), Roma AIB Sezione Lazio, 1982.
2. "Professione bibliotecario" *Bollettino d'informazioni - Associazione italiana biblioteche*, 28 (1988) No. 1/2.
3. Rosaria Campioni (ed.) *Fare biblioteca*. Percorsi formativi, professionalita e organizzazione del lavoro nelle biblioteche. Bologna, Analisi, 1988.
4. Attilio Mauro Caproni. *La formazione professionale del bibliotecario*. Milan, Editrice Bibliografica, 1989.

SHORT REPORT OF CONTINUING PROFESSIONAL EDUCATION IN POLAND

Maria Brykczynska
President
Warsaw Branch, Polish Library Assn.
Szczesliwicka 2 m 2, Warsaw 02-352, Poland

At present, the following institutions and organizations are responsible for training librarians employed in different types of libraries in Poland.

University centers responsible for the organization of postgraduate courses in library and information science. The first centers were organized in Poznan in 1968 and in Warsaw in 1969. These courses give library qualifications for those library workers who have master's degrees from any department other than librarianship and information science. They may be graduates in history or literature. Recently these professionals have been required to obtain this diploma in postgraduate studies in library science in order to advance in a library career.

These postgraduate studies are organized as extramural studies, usually for three semesters, in nine universities and higher pedagogical colleges: Gdansk, Sosnowiec, Krakow/Jagiellonian University, Krakow/Higher Pedagogical College, Lublin, Lodz, Warsaw, Wroclaw, Kielce.

In the case of Warsaw, there are two programmes of postgraduate studies: 1) a general programme in information science; 2) a special programme in information science for teachers. Some other special programmes such as those for people interested in collections of old books are planned for the future.

In Wroclaw, there are also two different curricula for postgraduate studies: 1) a general curriculum in library science; 2) a curriculum for teacher-librarians: knowledge of the organization and activities of school libraries; 3) a curriculum concerned with knowledge of readership.

Special courses for the training of teacher-librarians. There is a three-level system of professional education for this group. This system is also connected with promotion in library careers. All the professional training for this group is supervised by the Central Institute for Teacher Training.

Different seminars, courses, workshops are also organized by the Polish Library Association. Some of these are designed for librarians who must pass an examination before a State Examination Committee to acquire a diploma and become qualified librarians.

Other seminars such as those given by the Warsaw Branch of the Polish Library Association are connected with the present problems of Polish librarianship as well as with new trends in library and information science. These can be considered as refresher courses, giving an opportunity to the members of the PLA to update their professional knowledge.

SOUTH PACIFIC

CONTINUING PROFESSIONAL EDUCATION IN PAPUA NEW GUINEA

John Evans
Acting Head
Department of Information Studies
University of Papua New Guinea
P.O. Box 320, University PO., Papua, New Guinea

Past activities have recently been reviewed (1) and after several years of neglect the issue of continuing professional education is once more an active one in this country. The moving of the library school from the Administrative College of Papua New Guinea to the University of Papua New Guinea to a new Department of Library and Information Studies (2) will enable more to be done. The University has a greater scope for creating links with other institutions which can assist us in developing programmes.

It is expected that most courses will be run by the Department but on a collaborative basis making the best use of the limited professional staff available. The Department is already at an advanced stage of negotiation with a funding agency that will allow a planned programme of workshops to be provided in 1990, 1991 and 1992. In addition to this there is a possibility of assistance from another source to provide Australian experts to teach on the continuing education programme.

During 1989 only the following has been possible:

- Public Libraries Workshop (one week) provided by the National Library Service for the librarians of the various PNG provinces.
- National High School In-Service Workshop (one week) for librarians at the National High Schools. This was run at Sogeri National High School and supported by the PNG Department of Education. The School Library Service of the National Library Service also provides workshops for school librarians and teachers at various venues in Papua New Guinea.
- Archives and Records Management Workshop (three days). This was run by the University of Papua New Guinea Department of Library and Information Studies with assistance from the National Archives Service. This workshop inaugurates a series of further workshops to be provided during 1990.
- Seminar on the Use of Mini-Micro CDS/ISIS Software for Library and Information Management (two weeks) sponsored by Unesco and held at the University of Papua New Guinea.

Although limited, our recent experience has shown considerable enthusiasm for continuing professional education and, given the new home for the library school and the promise of adequate resources to fund courses, a very useful programme will be available for at least the next three to five years. The profession here needs all the help it can get.

References

1. Evans, John. "Short courses in librarianship in Papua New Guinea." *International Library Review*, 20 (1988), pp. 435-449.
2. Evans, John. "Transferred function: the moving of Papua New Guinea's Library School." *Education for Librarianship - Australia.* (1989) pp. 13-19.

CONTINUING LIBRARY EDUCATION IN NEW ZEALAND

Alan D. Richardson
Senior Lecturer in Librarianship
Victoria University of Wellington, Wellington 871, New Zealand

In the middle of 1988, it was beginning to look as if a planned programme of continuing professional education activities throughout New Zealand and coordinated by the New Zealand Library Association might have become a reality in 1989. To understand why this desirable state of affairs did not eventuate, and to appreciate the efforts made to overcome the difficulties needs some understanding of the background work on CPE which has been done over the past two or three years. Much could be written, and one day should be written, but, in this present context, only a broad outline is relevant.

For many years, librarians in this country have recognized that a serious and continuing gap exists in the provision of CPE. One of the real difficulties was in finding a person or group actually willing to do more than say that something had to be done. Another hazy area was deciding what belonged in the first professional qualification, and what was within the realm of on-the-job training, and what was appropriate as CPE activities. These matters remain under discussion.

From time to time, maybe in conjunction with an annual NZLA Conference, some individual or group, having perceived a need would organize a very good seminar which was both well attended and much appreciated. It is quite wrong to suggest that absolutely nothing was done for librarians. The real need was for the proper planning of on-going courses which would add up to something really worthwhile.

Other opportunities for librarians to extend their professional skills could be taken up on an individual basis through general courses run by local educational institutions such as Polytechnics. This is particularly true of courses in computer keyboard skills or some aspects of management. In May, 1987, one of the members of the NZLA Continuing Education Committee saw fit to remind members that "one of the measures of professionals is to take responsibility for their own continuing education and not always to expect their employers to pay for it, or their professional association always to supply it.

Nevertheless, NZLA members were aware that more could and should be done, and a really concerted effort began in 1987 to make some positive beginnings. In that year, the NZLA Continuing Education Committee commissioned a report from Margaret Trask, Executive Director of the Australian Information Management Association (AIMA). The report was written after a workshop on CPE which was held in conjunction with the 1988 NZLA Conference in Hamilton, and it was presented to the Committee in March. The report contains sixteen recommendations, too many to list here, but the matters which have a bearing on what will be mentioned later are these:

1. that the NZLA accept a policy of a three-year plan for continuing education;
2. that continuing education facilitators be appointed and trained by NZLA; and,
3. that a continuing education officer be appointed part-time for two years to assist with getting the whole programme under way.

Probably the most important of these recommendations was the third one. This person would have been the contact for all the Branch facilitators and would provide the much needed co-ordination of the three-year plan. The continuing education officer was on the point of being appointed when it was realized that the NZLA's immediate financial prospects were such that the appointment could not go ahead.

This was a real blow, and it must have been tempting to the Continuing Education Committee to give away the whole idea in the short term, but it is to their credit that, in spite of this serious difficulty caused by the lack of this key person, they continued to do their best with what they had and to make a tentative beginning. Branch facilitators were appointed, although their training course was also canceled at the last moment through lack of finance. Some courses thus became possible, and, in the circumstances, that was a remarkable achievement.

An outline of a three-year plan was made available to the facilitators by the Continuing Education Committee along with some comments and suggestions for year one (1989) and a recommendation to begin organizing a local plan as soon as possible. The Auckland Branch took a lead in this, since the Continuing Education Committee was also located there, and other Branch facilitators were provided with an outline of a suggested programme.

There is no question at all that the lack of the proposed continuing education officer has been very serious indeed, but individuals have done their best to continue with the programme. Specific courses suggested to Branches for year one were on the reference interview, service to Maori and Pacific Island people, and the use of CD-ROM in libraries. Some courses on one or more of these topics were mounted by some Branches with considerable success. In addition, information has been shared throughout New Zealand on planning, costs, speakers, venues, etc. for courses. A foundation has been laid in some areas on which future courses can be built.

The Wellington Branch has also formed a very useful liaison with the Center for Continuing Education at Victoria University of Wellington. A refresher course on cataloging for librarians was organized using the administrative services of the Center. The course was so successful that it was repeated some weeks later. It now seems very likely that the Wellington Branch will continue to develop that relationship as further courses are provided in the future.

Planning for 1990, the second year of the three year plan, is underway. It remains extremely unlikely that the Continuing Education Officer will be able to be appointed, and the training course for facilitators does not seem likely to eventuate either. In addition, the Continuing Education Committee is due to move from Auckland to Wellington so that a new group of people will have to decide what should and should not be attempted. It is vital that the small amount of impetus gained in 1989 should not be lost in 1990, but the profession as a whole is so committed to CPE that that really does seem very unlikely. We may have seen, in 1989, a tentative beginning to better provision of continuing education courses for librarians. Many people will hope that it is so.

SOUTH AMERICA

CONTINUING EDUCATION IN BRAZIL

Tania Botelho
SQN 316 - E-202 ASA Norte
Brasilia-De-Brazil 70775, Brazil

Continuing Education (CE), also called permanent education, has been planned and implemented by various kinds of institutions which have also made studies of professional needs, carried out job analysis, developed teaching methods, and sponsored specific activities such as courses, training, seminars, congresses, conferences, participation in working groups, reading groups, technical visits, etc. The Brazilian situation has demonstrated that the institutions dealing with CE have had some difficulty in implementing their programmes, although results have been satisfactory. Professional associations, university programmes, government, and private enterprises have developed a large number of events.

The main institutions promoting CE activities are:

- professional associations; there are 23 and all of them offer courses of short and medium duration, promote seminars and meetings, and also distant teaching;
- in addition to the Brazilian Congress on Library Science and Documentation, there are the permanent commissions of the FEBAB (Brazilian Federation of Librarians Associations) that are active in CE such as the commissions on Geosciences, on School and Public Libraries, and on Agricultural Documentation. They also promote seminars, meetings, and regional congresses;
- The IBICT (Brazilian Institute of Scientific and Technological Information) has developed activities in CE. They have used the existing university framework and are developing courses in cooperation with the universities. Besides the courses, they have promoted meetings, round tables, and conferences with guest speakers;
- The Secretariat of Industrial Technology of the Brazilian Center for Enterprise Support (STI/CEBRAE) together with the universities is currently promoting specialized courses in technological information;
- The Ministry of Education (MEC) through the Secretariat for Graduate Education (SESU) is currently promoting the National Programme for University Libraries (PNBU) that offers a course of specialization in university librarianship. It also promotes a National Seminar on University Libraries (SNBU) in cooperation with the university community;
- Library Schools offer courses at the Master's and Doctoral level, as well as specialized courses. They cooperate with the professional associations and the government. The existing infrastructure in the Library Science Schools has facilitated interaction among institutions developing CE activities. From the nucleus of cooperation, some studies and research have been done in subjects such as professional requirements, labor market, job analysis, development of teaching methods, and curriculum development; and,
- recently a private enterprise has begun to develop CE activities. They cover library science as well as informatics and organizational methods.

The professional societies have made a great contribution to the development of CE activities in the field of publication. We have in the country different kinds of periodicals, journals and specialized literature, playing a new role in the book market at the national level.

Although there are in Brazil different mechanisms and activities that support CE, the development of the information professional is still very inadequate. Some of the factors that we believe have contributed to this are:

- the professional interest in CE needs to be stimulated; there is inertia in the great majority of professionals.
- CE activities are expensive and the salary level of librarians is not always sufficient to permit them to attend.
- The participation in CE activities is not always considered in an individual's record for purposes of promotion and salary increases.
- Announcements of CE activities are not very well structured and marketing is inadequate to stimulate participation. Furthermore, the results of research do not appear systematically in the literature due to the lack of communication and interest on the part of the editorial boards of the reviews, periodicals, and journals. In such a situation, it is very difficult to arrive at a state of the art except for the informal communication network.
- Library schools have only recently begun to cooperate in CE activities and in building joint programmes with professional associations and the government.
- There is a need for a coordinated infrastructure of CE activities that will facilitate the planning and programme implementation on a more rational basis, in order to achieve the following goals:
  - Identification of professional needs and requirements for the 1990s.
  - Job analysis and requirements.
  - Identification of lecturers.
  - Possible sponsors to promote CE activities.
  - Wide communication of all kinds of events.
  - Development of modern teaching methods and audiovisual resources and methods.
  - Joint efforts with financing agencies and sponsors for the promotion of CE activities.
  - Joint efforts with research agencies to stimulate the promotion of research and support activities in CE.

References

1. Cunha, M.B. "O desenvolvimento profissional e a educacao continuada." *Revista Biblioteconomia Brasilia.* 12 (2) jul/dez, 1984, pp. 149-156. Analysis of propositions and concepts in CE activities as responsibilities of participating institutions.
2. Nocetti, M.A. "A educacao continuada para bibliotecarios: revisao da literatura." In *Estudos Avancados em Biblioteconomia e Ciencia da Informacao*, Brasilia, ABDF, 1984, pp. 13-32. A very complete literature review from different countries including Brazil. The situation in Brazil has not changed greatly from the scenario we have today.
3. Gadotti, M.*A. educacao contra a educacao.* Rio de Janeiro. Paz e Terra, 1982. 130 p. Analysis of fundamental aspects of educational philosophy for CE activities.
4. We have done a series of interviews with the managerial board of professional associations, government institutions, and library schools in Brazil.

APPENDICES

APPENDIX A

CONTINUING PROFESSIONAL EDUCATION RESOURCES IN THE UNITED STATES

Thelma H. Tate
Reference Coordinator
Mabel Smith Douglass Library
Rutgers University, New Brunswick, New Jersey, USA

Broadbent and Grosser identified six areas of focus for continuing professional development of librarians, including:

- the use of variety and mixed learning modes;
- cooperative programme planning that involves employers, educational institutions, professional and other associations;
- formal courses that meet expressed needs for information and management technology; and,
- greater personal and institutional investment of time and money in continuing professional development (Broadbent and Grossner, 1987, p. 99-115).

Works cited in this bibliography will help librarians to address some aspects of these issues and will provide practical knowledge for educational empowerment.

The search for excellence in continuing professional education is impacted by such factors as the need for special competencies for effective strategic planning in the global community; decision-making expertise to build better services in times of financial constraints; and creative, innovative approaches to enhance expertise of those in the profession. Educational competencies are required to plan technological, managerial, and transborder access to information for literacy development. Formal and informal CPE programmes are essential for the fulfillment of expressed educational needs to better serve the worldwide community.

As professionals, we are motivated to take responsibility for and develop mechanisms to help us learn more as we provide the best possible service for our clientele. This bibliography identifies competency building materials for individualized or group learning. For example, the cassette tape collections of the selected programmes of the American Library Association Annual Conferences give state of the art knowledge, research, and recommendations for library services in the post-electronic era. As was indicated in a recent brochure by the American Library Association, "ALA is continuing education." Likewise, various associations in the library and other disciplines develop excellent products to help facilitate ongoing in-service training for information specialists. The private sector is producing packets designed for individualized instruction. professional clearinghouses and exchange centers make available a wide range of workbooks, guides, templates, actual documents that can be borrowed and adapted for use in all types of libraries and information centers. The Library Orientation Exchange Center (LOEX) in Michigan (USA) is an example of these resource centers. This bibliography represents a very brief selected list of resources that are available to support learning for the professional.

The ALA Yearbook of Library and Information Services 1989, A Review of Library Events 1988. Edited by Roger H. Parent. Volume 14. Chicago: American Library Association, 1989.

Authoritative articles on subjects of enduring and current interest, including such topics as information technology, standards, international relations, organizations, membership activities, and biographies.

Bowker Annual; Library and Book Trade Almanac. 14th Ed. 1989-90/ New York: Bowker, 1989.

A condensed summary of the profession, including reports, legislation and funding, education and salaries, research and statistics, reference information, and organizations.

Breivik, Patricia S. and R. Wedgeworth, eds. *Libraries and the Search for Excellence*. Metuchen: Scarecrow, 1988.

Focuses on bibliographic instruction as the key to solving patron bewilderment. Identifies student, faculty, and librarian resistance to innovations as the principal problems of libraries in the search for academic excellence.

Breivik, Patricia and E. G. Gee. *Information Literacy*... New York: Macmillan, 1989.

Describes possibilities for increasing the levels of literacy in populations.

Broadbent, Marianne and Kerry Grosser. "Continuing Professional Development of Special Library and Information Center Managers," *Journal of Education for Library and Information Science*. 28 (2) : 99-115, Fall 1987.

A study of eighty-five Melbourne based special librarians and information center managers to determine their needs, aspirations, and professional development activities. Outlines findings of the study and discusses the implications for organizations special library and information centers.

Clark, A. J. "Education and Training for Librarianship and Information Work: Annual Bibliography, 1987," *Education for Information* 6 (3) : 257-314, September 1988.

This second annual international bibliography covers works published in a given year. It is arranged by subject, author, and title, and has a table of contents for access by broad subject headings. The College of Librarianship Wales was used as a basis for collecting this bibliography.

Dou, Henri, et al. "Teaching Bibliometric Analysis and MS/DOS, 1987," *Education for Information* 6(4): 411-423, December 1988.1

Explains the successful use of MS/.DOS to download references for reports and home-made databases.

Frost, Carolyn G. and Joseph Becker, eds. "Special Issue: Information Access in the 21st Century." Papers from a Symposium on Policy Issues in Knowledge and Information Access on November 5-6, 1986 in Ann Arbor, Michigan. *Information Society*. 5 (1): 1-55, 1987.

Fulton, Tara Lynn. "Reference Librarianship: Sharing Our Knowledge With Technical Service Colleagues," *RQ* 27 (2); 210-219, Winter 1987.

Discusses in-house information-sharing between Reference and Technical Services to orient new staff, staff development, and to facilitate interdepartmental understanding and cooperation.

Harter, Stephen and Susan M. Jackson. "Optical Disc Systems in Libraries: Problems and Issues," *RQ* 27 (4): 516-527, Summer 1988.

Discusses optical disc reference sources in terms of databases, search characteristics, education and training requirement, staffing costs, and funding. Emphasizes problems and issues generated by the adoption of optical disc technologies to reference services, and suggests roles for reference librarian in this electronic age.

Hill, Janet Swan. "The Literature of Description and Cataloging 1988," *Library Resources and Technical Services.* 33 (3): 258-270, July 1989.

Discusses theories and assumptions of catalogs and cataloging rules, the function and structure of catalogs, authority work, cataloging as well as management topics such as backlogs, preparation for cataloging.

Kemper, Marilyn. *Networking: Choosing A LAN Path To Interconnection.* Metuchen: Scarecrow, 1987.

Discusses strategies for computer communications, networking trends and issues, design, concepts, architecture and protocols, and motivation for connection, as well as staff training and innovations.

"Library Information and Information Resource Management ..." *ASLIB, The Association for Information Management.* Special Conference Issue. 16 (2): 36-48, February 1988.

This issue contains a series of articles that are of special interest for practicing librarians: "Information Industry;" "Bringing Education for Information Workers Into the 1980s;" "New Information professionals;" "Libraries and Information Resource Management;" "On the Nature and Types of corporate Information Center;" "User Studies and Evaluation Services." Each monthly issue of ASLIB includes a section on Professional Development Highlights on timely topics, such as END USERS OR CD-ROMs. The Association of Information Management has an Information Resource Center that provides ongoing educational materials and services to librarians. The address is: ASLIB INFORMATION, Information House, 26-27 Boswell Street, London WC1N 3JZ.

LOEX Clearinghouse. Eastern Michigan University Library. Ypsilanti, Michigan 48197. (313) 487-0168.

The LOEX-Library Orientation Clearinghouse provides full borrowing privileges of sample instruction materials and information and referral services. A quarterly newsletter helps to keep librarians up to date on recent materials and conferences.

McCabe, Gerard B., ed. *The Smaller Academic Library, A Management Handbook.* New York: Greenwood, 1988.

A condensed guide that includes information about many aspects of libraries and services, such as administration, personnel, budget and finance, collection development, user programmes and services, technical services, and physical plant.

Martell, Charles. "Achieving High Performance in Library Work," *Library Trends,* 18 91): 73-91, Summer 1989.

Explores the concept of productivity in relation to the transforming nature of work. Discusses motivation, job design, quality of work life, organizational culture, high performance, excellence and renewal.

"The Mission of the Public Library," *Public Library Quarterly* 9(1): 35-38; 9(2); 37-40; 9(3): 31-36, 1989.

Shares Mission Statements from three libraries in the United States: the Public

Library in Alamedia, California; the Monroe County Library System in Monroe, Michigan; and the County of Henrico Public Library in Virginia.

Mount, Ellis, ed. "One Hundred Years of Sci-Tech Libraries: A Brief History," *Science and Technology Libraries* 8(1): 1-88, Fall, 1987.

Discussion thirty-five years of information retrieval in academic science and technical, government, corporate libraries in the 19th-20th centuries.

Office of Management Services, Association of Research Libraries. 1527 New Hampshire Avenue, N.W., Washington, D.C., 20036. (202) 232-8656.

The Systems and Procedures Exchange Center of the Association of Research Libraries is "an information clearinghouse to help libraries learn of current practices, solve problems, manage change, and improve performance." Each year the Systems and Procedures Exchange Center of ARL develops and makes available ten SPEC KITS with FLYERS. Each kit provides both some theoretical background and some very practical education to update librarians expertise and skills. Also, each FLYER includes techniques and procedures used in some specific libraries in the United States. The SPEC Kits cost twenty dollars each. In 1989, the following SPEC KITS were made available: 1) *Qualitative Collection Analysis: The Conspectus Methodology.* #151 February 1989; 2) *Brittle Books Program.* #152. March 1989; 3) *Use of Management Statistics.* #153 April 1989; 4) *Computerized Online Bibliographic Searching in ARL Libraries.* #154. May 1989; 5) *Onsite Access to Library Collections by Visiting Scholars.* #155. June 1989; 6) *Automated Authority Control in ARL Libraries.* #156. July/August 1989.

"proceedings of the 1988 CLA Convention, A Selection of Presentations ..." *Catholic Library World* 60 (1): 5-37, July/August 1988.

Katherine Paterson, a noted author of children's books, Gordon C. Zahn, and John Wilcox were among the presenters included in the selections.

Prytherch, Ray. *Staff Training in Libraries, The British Experience.* England: Gower, 1986.

Addresses state of the art issues of economic constraints and training in public, special, academic, governmental, school libraries in Great Britain. Provides an overview of cooperative training, field work, and training of British librarians. The complimentary volume entitled *Handbook of Library Training practice* by the same editor and publisher (1986) provides a blueprint for good library practice, including pervasive techniques, user-staff relationships, and training in specific skills.

Sellberg, Roxanne. "The Teaching of Cataloging in U.S. Library Schools," *Library Resources and Technical Services.* 32 (2): 30-47, January 1988.

Discusses the gap between knowledge needed by new catalogers and knowledge gained in library school. Includes implications for future curriculum development.

Smith, Kent A. and Patricia E. Healy. "Transborder Data Flows: The Transfer of Medical and Other Scientific Information by the U.S.," *Information Society* 5(2): 67-75.

Provides an overview of cultural, political, economical, social, and policy issues. Includes experiences of the U.S. National Library of Medicine pertaining to transfer of biomedical information worldwide.

Stone, Elizabeth W. and Jean L. Preer. "Empowerment For Libraries and Literacy: A General Federation of Women's Club Tradition," *Public Libraries* 28 (5): 298-299, September/October 1989.

Highlights the history and partnership of GFWC and libraries to promote literacy. Demonstrates the roles women have played historically to enrich the population.

Svinicki, Marilla D. and Barbara A. Schwartz. *Designing Instruction for Library Users, A Practical Guide*. New York: Marcel Dekker, 1988.

An academic librarian's guide to development of objectives and planning for teaching process. Sequencing, theory, case studies and samples of scales, questions, workbooks and guides.

Tifft, Rosamond. "The Growth and Development of Information and Retrieval in Library Services: A Selective History and Review of Some Recent Developments," *The Reference Librarian* 21: 229-259, 1988.

Discusses the meaning of the terms "information" and "retrieval in public libraries since 1960.

Weatherford, John H. *librarians 'Agreements: Bargainingfor a Heterogeneous Profession.* Metuchen: Scarecrow, 1988.

Reports on fifteen years of collective bargaining and includes agreements from Boston Public Library, University of California Federation of Librarians university Council, and State University System of Florida and Unified Faculty of Florida.

Weech, Terry L. "national Accreditation of Public Libraries: A Historical Perspective,"- *Public Libraries.* 28 (2): 119-125, March/April, 1989.

An historical review of proposals made between 1961-1988 for the accreditation of public libraries nationwide. Barriers to national accreditation of public libraries are discussed and the prospects for success of the Commission on the Accreditation of Public Libraries proposals are weighed.

White, Herbert S. *Librarians and the Awakening from Innocence, A Collection of Papers.* Massachusetts: G. K. Hall, 1989.

Discusses and analyses various aspects of library education, library and training management, interrelationship between education and practice, economic and costs, users and librarians, as well as political processes.

Williams, Robert V. "Using the Information Resources of the Global Village: The Information Systems of International Intergovernmental Organization," *Special Libraries* 80 (1): 1-8, Winter 1989.

Contains significant information resources of interest to librarians worldwide. Discusses the nature of the information systems of IGOs in general with particular attention to problems of acquiring information from them.

APPENDIX B

CUSTOMER COMPUTER SEARCH ON DISTANCE EDUCATION
(An ERIC Clearinghouse on Higher Education Search)

Some descriptors used in this search included:
Higher education
Postsecondary education
College
University
Distance education
Communications media
Nontraditional education
Extension education
Foreign countries
Canada
United Kingdom
Britain
Iran
Australia

Alnaes, T. "Library and Information Science Education in Norway." *Education for Information*, v. 6, n. 2, pp. 187-97, June 1988.

Arblaster, John R. "Contact North: The Concept, Policy, Development, and Status of the Northern Ontario Distance Education Access Network." 30 p., September 1988. Paper presented at the Annual Conference of the Association of Canadian Community Colleges (Saint John, New Brunswick, May 30-June 1, 1988).

Bowles, John C. "Computerized Educational Delivery Strategies in Nine North American Colleges." *Programmed Learning and Educational Technology*, v. 25, n. 1, pp. 34-45, February 1988.

Bray, Catherine. "Women's Studies at a Distance: Experiences of Students and Tutor." *Canadian Journal of UniversityContinuing Education*, v. 14, n. 2, pp. 37-49, Fall 1988.

Carl, Diana R.; Densmore, Bruce. "Introductory Accounting on Distance University Education via Television (DUET); A Comparative Evaluation." *Canadian Journal of Educational Communication*, v. 17, n. 2, pp. 81-94, Spring 1988.

East, James R. "Teaching on Weekends and in Shopping Centers. A Guide for Colleges and Universities." 179 p., 1988. Available from: Indiana University-Purdue University at Indianapolis Bookstores, Campus Services Building No. 4, 1830 W. 16th Street, Indianapolis, IN 46202 ($25.00)

Graff, Kurt, ed.; Holmberg, Borje, ed. "International Study on Distance Education: A Project Report." *FernUniversitat, Hagen (West Germany). Zentrales Inst. fur FernstudienforschungArbeits-bereich.* 151 p., November 1988.

Hawkridge, David. "Distance Education and the World Bank." *British Journal of Educational Technology*, v. 19, n. 2, pp. 84-95, May 1988.

Innes, Sheila. "Ensuring Quality in the Open College." *Education + Training*, v. 30, n. 3, pp. 10-11, May-June 1988.

Kinser, W.; Pear, J. J. "Computer-Aided Personalized System of Instruction for the Virtual Classroom." *Canadian Journal of Educational Communication*, v. .17, n. 1, pp. 21-36, Winter 1988.

Kirschner, P.A.; Meester, M. A. M. "The Laboratory in Higher Science Education: Problems, Premises and Objectives." *Higher Education*, v. 17, n. 1., pp. 81-98, 1988.

McIsaac, Marina Stock, et al. "Examining Distance Education in Turkey." *Distance Education*, v. 9, n. 1, pp. 106-14, March 1988.

McIsaac, Marina Stock; et al. "Video Education Centers To Meet Student Needs in Turkish Distance Education Programmes." 3 p. Paper presented at the Annual Meeting of the Association for Educational Communications and Technology (New Orleans, LA, January 16-21, 1988).

McLellan, James L., ed; Taylor, William H., ed. "Canadian Association for the Study of Adult Education Proceedings of the Annual Conference (7th, Calgary, Alberta, Canada, May 1988)." May 1988, 346 p.

"Management Education--Future Needs: 1" *Education + Training*, v. 30, n.2, pp. 9-10, March-April 1988.

Mason, Robin. "Computer Conferencing: A Contribution to Self-Directed Learning." *British Journal of Educational Technology*, v. 19, n.1, pp. 28-41, January 1988.

Misanchuk, Earl R.; Brack, Robert E. "A Method for Setting Priorities for Programming in University Extension." *Canadian Journal of University Continuing Education*, v. 14, n. 1, pp. 60-73, Spring 1988.

Nettlefold, Brian A. "Library Services for Off-Campus Students: A Study of Canadian Practice and Its Potential for ireland." National Inst. for Higher Education, Dublin (Ireland). 32 p. 1988.

Noll, Cheryl L.; Emerson, Tricia M. "Adult Education in the Soviet Union." 36 p. September 28, 1988.

Paulet, Robert O. "Factors Influencing Successful Counseling in Selected Distance Education Programmes." *Journal of Research and Development in Education*. v. 21, n. 3, pp. 60-64, Spring 1988.

Paulet, Robert O. "Special Education, Distance Education and Minority Cultures." *History and Social Science Teacher*, v. 23, n. 4, pp. 207-11, Summer 1988.

Pear, Joseph J., Kinsner, W. "Computer-Aided Personalized System of Instruction: An Effective and Economical Method for Short- and Long-Distance Education." *Machine-Mediated Learning*, v. 2, n. 3, pp. 213-37, 1988.

Rumble, Greville. "The Use of Microcomputers in Distance Teaching Systems. ZIFF Papiere 70." *FernUniversitat. Hagen (West Germany). Zentrales Inst. fur Fernstudienforschung Arbeitsbereich.* 27 p. March 1988.

Scott, R. Neil. "The University of Oxford's Approach to Adult and Continuing Education." *Journal of Academic Librarianship*, v. 14, n. 1, pp. 15-23, March 1988.

Slade, Alexander L.; Webb, Barbara. "The Canadian Off-Campus Library Services Survey, 1988." 13 p. January 1988.

Straesser, Rudolf. "Mathematics in Adult, Technical and Vocational Education Including Distant Education. Pre-Conference paper for Action Group 7 of the International Congress on Mathematics Education (6th, Budapest, Hungary, July 27-August 3, 1988). Occasional paper 97." 14 p. March 1988.

Tresman, Susan, et al. "The Potentiality of Distance Learning." pp. 687-91, June 1988.

Williams, Sylvia; Sharma, Pramod. "Language Acquisition by Distance Education: An Australian Survey." *Distance Education*, v. 9, n. 1, pp. 127-46, March 1988.

Worthy, Ward. "Montreal Trio Brings Chemistry to the Public." *Chemical and Engineering News*, v. 66, n. 25, pp. 29-31, June 20, 1988.

Wright, Clayton R. "Independent Study Course Development Costs." *Canadian Journal of Educational Communication*, v. 17, n. 2, pp. 95-105, Spring 1988.

Zahlan, Anthony B. "Issues of Quality and Relevance in Distance-Teaching Materials." *Prospects*, v. 18, n.1, pp. 75-83, 1988.

APPENDIX C

SCOLE Guidelines for Quality in Continuing Education For Information, Library and Media Personnel*

CONTENTS

* Continuing Education Subcommittee of the Standing Committee on Library Education (SCOLE);
Adopted by the ALA Council on January 12, 1988;
American Library Association, 50 East Huron Street, Chicago, Illinois 60611, USA.

Forward

The Programme for Quality in Continuing Education for Information, Library, and Media Personnel was established as part of a research and development project funded by the Office of Education in conjunction with CLENE, Inc. [the Continuing Library Education Network and Exchange]. The Programme consisted of 1) a policy statement specifying the definition, goals, and benefits of continuing education; 2) criteria for quality continuing education; and 3) a system for the approval of providers of continuing education.

Since CLENE has subsequently become a round table of the American Library Association [CLENE RT], the ALA Standing Committee on Library Education [SCOLE] has formed a continuing education subcommittee, one of the functions of which is to transform the Criteria into ALA Guidelines.

This document represents the SCOLE Continuing Education Subcommittee's effort to adapt the Criteria to the ALA Guidelines format.

The Continuing Education Subcommittee wishes to acknowledge the efforts of the members of the CLENE Council on Quality Continuing Library Education and the staff of the project who developed the original *Criteria for Quality:* William Asp, Bernard Franckowiak, John Hinkle, John G. Lorenz, Margaret Myers, Debra Nygaard, Ruth J. Patrick, Gary Purcell, Elizabeth W. Stone, Alphonse F. Trezza, Susan Viola, and Micki Jo Young.

The following individuals worked on the revision of the document to adapt it to the ALA Guidelines format:

ALA Continuing Education Subcommittee of SCOLE

Darlene E. Weingand, Chairperson (1986-1987)
Mary C. Chobot, Chairperson (1987-1989)
Margaret E. Myers, Staff Liaison
Thelma Ananias Morris
Barbara Newmark-Kruger
Richard J. Sorensen
Nancy Starke
Jana Varlejs
Mary Y. Moore, CLENE RT Liaison

Introduction

Continuing Education

Continuing education is a learning process which builds on and updates previously acquired knowledge, skills, and attitudes of the individual. Continuing education comes after the preparatory education necessary for involvement in or with information, library, media services. It is usually self-initiated learning in which individuals assume responsibility for their own development and for fulfilling their need to earn. It is broader than staff development which is usually initiated by an organization for the growth of its own human resources.

Goals and benefits

Continuous learning is an obligation of all those who work to identify, select, organize, retrieve, disseminate, and make accessible the record of human thought and actions. The goal of continuing education is to improve information, library, media services by maintaining or improving competence of practitioners. Regular participation in continuing education activities enables the practitioners to:

- refresh basic education by mastering new concepts in a constantly changing environment;
- keep up with the new knowledge and skills required to perform their roles responsibly;
- prepare for specialization in a new area; and,
- enjoy the intrinsic satisfaction resulting from learning.

Scope

A wide variety of learning opportunities is available both on the job and outside the workplace. These opportunities include both formal and informal learning activities and exist independently or under the guidance and sponsorship of various institutions, associations, and other organizations. Appropriate subject areas are defined broadly to include all related disciplines that help participants to practice with competence, thereby granting individuals wide latitude in selecting subject matter and format appropriate to their present or future needs. Although individuals assume primary responsibility for their own development, continuing education encompasses shared responsibility, cooperation, and interaction among a number of library and educational agencies, relevant federal agencies, the individual's place of employment, and a wide variety of professional and other educational associations at the local, state, regional, national, and international levels.

Need for assurance of quality

Due to the rapid changes occurring throughout the information, library, media fields, and indeed, the society they serve, people have sought means to keep abreast of the new developments. This demand was met by a supply of numerous and various continuing education activities from many sources.

In quality and effectiveness, they have ranged from excellent to less than adequate. Insufficient attention and inadequate experience were being brought to the quality of the activities and there was a lack of coordination.

The members of the information, library, media community suffered the consequences of this unevenness and asked that something be done to upgrade the quality of the programmes and to institute a form of accountability to assure that the programmes justify the time, energy, and costs expended by the participants, by their employers, and by the providers themselves.

Before CLENE (the Continuing Library Education Network and Exchange) became an ALA round table, it conducted a research and development project funded by the Office of Education to create The Programme for Quality in Continuing Education for Information, Library and Media Personnel. The Programme consisted of 1) a policy statement specifying the definition, goals, and benefits of continuing education; 2) Criteria for Quality continuing education; and 3) a system for the approval of providers of continuing education, which was administered by the National Council on Quality Continuing Education for Information, Library, Media Personnel [see Appendix A].

Now that CLENE RT is officially affiliated with ALA, and the former Programme, through the National Council, continues to use the original criteria, there is a need for a formal ALA document which establishes guidelines for quality in continuing education. In 1983, Elizabeth Stone published in the ALA Yearbook a survey which also demonstrated a concern for quality in continuing education. In addition, recent hearings (SCOLE/Library Education Assembly/CLENE RT Hearing on July 6, 1985, in Chicago) addressed the question "Whither ALA in Continuing Education" and testimony referred to the ALA policies on continuing education which contain a statement on the need to have a strong relationship between ALA's continuing education programme and Association policies. The development of guidelines would provide the policy statements with respond to these stated needs.

If ALA, as the primary and largest professional association, is to provide direction for profession-wide continuing education, there must be in place some formal written documentation which speaks to the critical issue of quality. The development of guidelines is a major step forward.

Criteria for Quality

The delivery of quality continuing education programmes appropriate to the needs of people involved in information, library, media services must be based on careful planning. A number of factors should be used in assessing the planning, design, delivery, and evaluation of these programmes. These criteria can be applied by any person or group interested in providing quality continuing education opportunities.

Criteria are somewhat different for each type of continuing education opportunity. Criteria follow for five different kinds of continuing education strategies: Group Programmes and Activities; Individualized programmes and Activities; Instructional Materials and Technologies; Continuing Education Providers; and Learning Consultants for Individualized Continuing Education Programmes.

Group Programmes and Activities

Definition: A group programme or activity is one where the participant learns a given subject through interaction with an instructor(s) and other participants. This includes such gatherings as academic courses, institutes, workshops, seminars and short courses.

Needs Assessment. Criterion: The specific needs of the client group should be assessed.

Assessment Factors:
1. has the target group been clearly defined?
2. Has a systematic needs assessment been conducted?
3. Have representatives of the target group participated actively in the needs assessment process?

Objectives. Criterion: Specific and measurable/observable learning objectives should be stated in one or more of the following areas: changes in attitude and approach to the solution of problems; acquisition (or mastery) of new knowledge or the revision of outdated knowledge in specific skills, techniques, and procedures.

Assessment Factors:
1. Are the objectives based on the needs assessment?
2. Do the objectives describe the expected outcomes for the participants?
3. Are the objectives measurable/observable?
4. Do the objectives address changes in the participant's work performance as a basis for evaluating impact?

Design and Presentation. Criterion: The programme should be systematically designed and delivered to meet the statement of objectives.

Assessment Factors:
1. Are instructional strategies, i.e., learning experiences, teaching methods, and content materials appropriate and effectively employed in achieving the objectives?
2. Are person(s) knowledgeable in the content area as well as instructional design techniques involved in planning and conducting the educational offering?
3. Does the design consider the participant's entry level competency?
4. Is sufficient time allotted for the participant to meet the objectives?
5. Is consideration given to individualization, such as learning styles, pacing, and assimilation rate?
6. Are principles of adult education, including involvement, self-direction, and reinforcement included in the design of the programme?

Timeliness. Criterion: The programme content should be current and timely.

Assessment Factors:
1. Are the concepts presented up-to-date with the current ideas, trends, facts, etc.?
2. Is the programme content reviewed periodically to assure it is accurate and consistent with currently accepted practices and techniques?

Promotion. Criterion: The educational offering should be promoted responsibly.

Assessment Factors:
1. Is the promotion directed at the audience that may best benefit from the activity?
2. Does the promotional material include the following:
   - the scope of the subject content;
   - a statement of the objectives in terms of learner outcomes;
   - the nature of the target audience;
   - the level of the offering, i.e., beginning, intermediate, advanced, or overview;
   - the location, time, and schedule of activities;
   - the name(s) and qualifications of the presenter(s);
   - the cost, the items covered by the cost, and provisions for refund;
   - the amount and kind of credit being offered, if applicable;
   - the preparation, if necessary, which participants should engage in prior to the programme;
   - a description of the learning activities, methods, and resources used in the programme?

Evaluation. Criterion: Evaluation should be an on-going and integral part of the educational offering.

Assessment Factors:
1. Is the evaluation based on the stated objectives?
2. Does the evaluation cover the starting point, the process, and the outcomes of the learning experience?
3. Do the evaluation procedures include provision for adjusting the programme in response to new needs identified during the course of the programme?
4. Does the evaluation assess the appropriateness of the content as well as the effectiveness of the materials and techniques employed?
5. Does the evaluation include the performance of the presenters and staff?

.6. Does the evaluation include adequacy and appropriateness of the facilities?
.7. Where confidential or sensitive materials or results are involved, does the evaluation process take into consideration ethical handling of these items?
.8. Is outside assistance sought as needed [e.g. for data analysis?]
.9. Are the evaluation results made known to the participants?
10..Are the evaluation results used to improve future programmes?
11..Does the evaluation process clearly indicate that the participant has the right .to express grievances or report unethical procedures?

Individualized programmes and activities

The criteria for Group Programmes and Activities (p. 146) and Instructional Materials (p. 149) are applicable to individualized continuing education learning activities. In applying the Criteria to the design and accomplishment of an individualized continuing education activity, special attention must be given to the following additional points:

Needs Assessment. Criterion: The individual's specific learning needs or interests have been identified and assessed.

Assessment Factors:
1. Has an appropriate needs assessment process been followed?
2. What are the needs or interests based upon, e.g. job, career growth, personal interest, etc.?
3. Are colleagues consulted or outside resources utilized in determining needs?

Objectives. Criterion: Based on the needs assessment, a set of clear, measurable, and/or observable learning objectives have been stated in one or more of the following areas: changes in attitudes; updating of outdated knowledge and/or attainment of new knowledge; and awareness or mastery of specific skills and techniques.

Assessment Factors:
1. Are the objectives clearly stated?
2. Are the objectives appropriate in meeting the assessed needs?
3. Will the objectives contribute to the competence of the individual?
4. Are the objectives reasonable and attainable given the individual's current expertise and the available resources and time?
5. Do the objectives specify the expected outcome for the individual in terms of the level of knowledge or performance that is expected to be attained?
6. Are the objectives stated in measurable terms so that they can be evaluated? How will the individual know or be able to demonstrate achievement of each objective?
7. Do the objectives address changes in future role performance of the individual as a basis for evaluating impact?

Learning plan. Criterion: An organized and coherent series of learning activities is systematically planned to achieve the individual's stated objectives.

Assessment Factors:
1. Are learning activities appropriate to the objectives?
2. Do the activities utilize appropriate materials, resources, or persons expert in the content/skill area?

3. Are learning activities organized, sequenced, and paced so as to be manageable and to build effectively upon each other?
4. Are appropriate and effective educational strategies/techniques employed that respond to the individual's learning style?
5. Are the activities at the appropriate level for the individual's level of competence?
6. Is adequate time allocated to achievement of the objectives of each activity?

Evaluation. Criterion: Evaluation is an on-going and integral part of the individualized continuing education activity.

Assessment Factors:
1. Is the evaluation based on the stated objectives?
2. Has the plan of activities been followed?
3. Does the evaluation cover the starting point, the process, and the outcomes of the learning experience?
4. Does monitoring provide opportunity for reassessment of objectives and adjustment of the programme in light of newly realized needs/interests or unrealistic expectations?
5. Are resources and experts utilized as planned?
6. Do the monitoring and evaluation look at appropriateness of the content and resources and the effectiveness of learning strategies?
7. Do monitoring and evaluation effectively assess the individual's level of achievement of the objectives and provide opportunity for corrective efforts?
8. Does the evaluation encourage application of learning to the work situation?
9. Is the monitoring and evaluation process carried out with appropriate review, assessment, or consultation with a qualified, unbiased person?

Instructional materials and technologies

Definition: Instructional materials and technologies are terms applied to the broad area of educational techniques that go beyond the live instructor speaking to a group of learners. This includes: 1). materials used by the instructor in the group programme or activity; 2). delivery of programmes through telecommunications (e.g., open and closed circuit television or broadcasts by radio or satellite); and 3). self-instructional materials for individual learners (e.g., programmed texts, multimedia kits, and computer assisted instructional programmes).

All instructional materials and technologies must be designed, developed, and/or utilized in a manner consistent with the tenets for quality in all educational programmes. Some types require special considerations due to their nature.

Suitability. Criterion: The instructional materials and technologies should be suited for the learning audience.

Assessment Factors:
1. Does the material specify any necessary prerequisites?
2. Is there a list of terms or jargon that must be known before the instructional materials can be used effectively?
3. Is the stated level of the instructional material the same as the learning audience, e.g., beginner or novice, moderate level or expert?
4. Is the delivery system appropriate to the learning activity?
5. Is the technical quality of the delivery system adequate?

6. Have materials and technologies selected for the learning activities been selected on the basis of sound evaluation methods?

Objectives. Criterion: A set of clear measurable/observable learning objectives has been stated in one or more of the following areas: changes in attitude; attainment of knowledge; and awareness or mastery of specific skills and techniques.

Assessment Factors:
1. Are the objectives measurable/observable?
2. Do the objectives specify the expected outcome for the individual in terms of the level of knowledge, attitude, or performance that is expected to be attained?

Learning strategies. Criterion: The instructional materials utilized are suitable and appropriate for the objectives of the learning activity.

Assessment Factors:
1. Are types of media used to accommodate different learning styles and preferences?
2. Is a variety of media used to stimulate the participant's interest and motivation?
3. Are the instructional materials suitable for the chosen mode of instruction, i.e., large-group, small-group, or independent study?
4. Are the instructional materials suitable for a variety of purposes: to set a mood or foster an attitude, to present facts, to illustrate or augment existing facts, to summarize a topic, or to evaluate or test learning?

Timeliness. Criterion: The instructional materials are current and timely.

Assessment Factors:
1. Are the concepts presented up-to-date with the current ideas, trends, facts, etc.?
2. Are the illustrations timely:
   a) appropriate to the historical period covered, or
   b) reflective of today's culture, styles and equipment?
3. Is the production date or revision date given?

Learning environment. Criterion: The instructional materials are utilized in an appropriate facility and learning environment.

Assessment Factors:
1. Is the size of the room appropriate?
2. Is the lighting adequate?
3. Is the layout of the room appropriate?
4. Is the room equipped with the necessary electrical wiring, sound system, screen(s), and furniture to accommodate the equipment used with the instructional materials?

Package or module design. Criterion: An instructional package (or module), whether produced or purchased, is a complete and integrated unit and is well documented.

Assessment Factors:
1. Is there a statement of objectives for the material?
2. Is the target audience identified?
3. Are the credentials of the subject matter experts provided and are they satisfactory?

4. Is evidence supplied that the producer tested and revised the material? Are the characteristics of the trial group sufficiently similar to this group of participants?
5. Is the technical quality adequate so there is good clarity, accurate exposure, synchronization, no distortion, etc.?
6. Does the presentation of the content match the attributes of the medium?
7. Is there an instructor's kit which includes:
   a) an instructor's guide;
   b) questions for discussion;
   c) additional problems or exercises;
   d) materials for display or demonstration;
   e) visuals/audio materials/software used in the programme; and,
   f) additional or supplemental materials for distribution?
8. Is all equipment needed to utilize the instructional materials standard, or, if not, are specific hardware specifications given?
9. If developed as a self-instructional package, is there a learner's kit which includes:
   a) a set of instructions;
   b) a list of learning goals and objectives;
   c) appropriate software;
   d) a workbook or other types of problems or exercises;
   e) visuals/audio materials/software used in the programme;
   f) evaluation instruments, both interim and final;
   g) additional or supplemental material; and,
   h) suggestions for enrichment or further study?

Continuing education providers

Organizations or individuals providing continuing education should use the following criteria and assessment factors to assess their capability to design and deliver appropriate educational experiences.

Administration. Criterion: A structure for systematic administration of the continuing education programmes should be established.

Assessment Factors:
1. Is one individual assigned the ultimate responsibility for administering the programme?
2. Is the assigned individual qualified to carry out his/her responsibility by virtue of background and experience?
3. Is there continuity of administration?
4. Is there evidence of careful, long range planning for the overall programme?
5. Is there a recognition system (such as the awarding of CEUs) so that student participation can be documented?

Human Resources. Criterion: continuing education presenters, designers, and other human resources should demonstrate current expertise by training and experience in the pertinent subject matter and/or educational methodologies as appropriate to their function.

Assessment Factors:
1. Are the presenters and human resources adequate in number for the type of learning method used?

2. Do the presenters have expertise, not only in subject matter, but in setting educational objectives, teaching, and assessment of educational results of the programme?
3. Do the presenters have the ability to facilitate learning and communication effectively?
4. Are appropriate adult education methods used in the programme?

Facilities. Criterion: Adequate facilities conducive to good learning should be available for continuing education programmes.

Assessment Factors:
1. Is there appropriate and sufficient space for the educational methods used, including learner participation?
2. Do facilities include pertinent and adequate library resources, space for conferences, audiovisual equipment and other resources as appropriate?

Financial Resources. Criterion: The organization's financial resources should be adequate to support the educational programme.

Assessment Factors:
1. Is there evidence of sound fiscal planning and management, e.g., a clearly identifiable component of the total budget earmarked for continuing education?
2. Does the budget include (as needed) the following:
   a) clerical and secretarial services;
   b) audiovisual and other pertinent equipment;
   c) supplies and promotional materials;
   d) travel and other related costs;
   e) other costs as appropriate?
3. Have financial resources to augment participant fees been sought to support the programme, if appropriate?

Learning consultants for individualized continuing education programmes

Individuals may pursue learning activities by themselves, or with the assistance of a learning consultant.

Selection of a learning consultant. Criterion: Persons wishing to work with learning consultants for individualized continuing education activities should consider the following assessment factors in making their selection.

Assessment Factors:
1. Can the learning consultant demonstrate competence in planning and evaluating education programmes, especially those involving individualized learning?
2. Does the learning consultant have knowledge of the specialty area being pursued by the learner or have access to persons with that knowledge?
3. Has the learning consultant developed a plan that states:
   a) the content area and level of training he or she is qualified to supervise;
   b) the kind of advisor services he/she is prepared to give;
   c) the general model that will be followed in the assessment of learning needs ....and the development, monitoring, and evaluation of activities?

Assistance. Persons wishing assistance in identifying a learning consultant may contact the ALA Office for Library Personnel Resources or the CLENE Round Table.

GLOSSARY

Activity - An individual educational experience or offering, such as a lecture, Home Study course, workshop, seminar, symposium, etc; a discrete but coordinated component of a programme designed to produce results that help to achieve programme objectives as well as more specific education objectives.

Adult learning - The study of how adults learn compared to how youth learn. Classic works in the field are done by Knowles, Houle, and Knox.

Colloquium - A series of lectures or seminars convened by a coordinator, but featuring different speakers or leaders.

Competency - Ability, skill or fitness based on meeting certain minimum requirements.

Conference - A general type of meeting usually of one or more days' duration, attended by a fairly large number of people. A conference will have a central theme but is often loosely structured to cover a wide range of topics. The emphasis is on prepared presentations by authoritative speakers, although division into small group sessions for discussion purposes is often a related activity.

Consultant - A qualified individual who gives advice or services regarding matters in his or her area of specialized knowledge or training.

Contact Hour - A standard unit of measurement sometimes equated to 50 minutes of participation in an organized learning activity.

Continuing Education - Planned learning experiences utilized by individuals following their preparatory education necessary for entrance into the field. Continuing education is a generic term which includes staff development as one of its elements. Similarly, in-service training and orientation are subsets under staff development. Whereas continuing education takes as its base the individual, staff development uses as its base the development of the group as its relates to the total organizational system. Continuing education opportunities include both formal and informal learning situations, and need not be limited to library subjects or the offerings of information science schools.

Continuing Education Unit (CEU) - Ten contact hours of participation in an organized continuing education experience under responsible sponsorship, capable direction, and qualified instruction.

Correspondence Course - A formal, independent study programme consisting of reading and exercises submitted to a knowledgeable instructor for evaluation. The course usually covers several lessons to be completed in a specified period.

Course - A programme of instruction as in a college or university; a series of sessions prepared for a group by the instructor usually with a well-defined subject and scope. Commonly focused on the dissemination of knowledge, but sometimes may include skills building for attitudinal training.

Criterion: A standard by which a programme, activity, etc. can be judged or evaluated.

Educational Technology- The development, application, and evaluation of systems, techniques and aids to improve the process of human learning.

Educator/Educational Staff - Individuals with the prime responsibility for the conduct of the learning experience; those who enable learning through skilled and deliberate efforts, those agents of learning upon whom so much of the success of each activity will depend. Synonymous terms include teachers, instructors, faculty, facilitators, leaders, trainers, consultants, and resource people.

Evaluation- Systematic investigation of the worth or merit of a project, programme, activity, instructional material, etc.

Formative Evaluation - Evaluation designed and used to improve a project, programme, activity, etc., designed to present conclusions about its merit or worth and recommendations about whether it should be retained, altered, or eliminated.

Field Work (see **Practicum**)

Goal - Purpose; an intended direction or aim; an ultimate outcome; an overall definition of what a programme or activity is to accomplish.

Hardware- A generic term for audiovisual and computer equipment.

Individualized Method of Learning- A method of learning which leads the learner to control the rate of his/her progress in learning. The communication medium is correspondence (mail), the computer, programmed text, or teaching machine, independent learning on campus.

Institute - Generally similar to a conference, but more tightly structured to provide a more systematic development of its theme, with the emphasis more on providing instruction in principles and techniques than on general information. Participants are usually individuals who already have some competence in the field of interest. Institute programmes may have certain continuity, meeting on a yearly basis for example.

Instructional Aid - Any item (hardware, software or printed materials) used to support instruction.

Instructional Materials - Any print or audiovisual resource which supports instruction. [for example, handouts, slides, transparencies, etc.].

Instructional Technologies- Hardware, software and/or telecommunications which are used to assist the learner in engaging in an educational activity.

Internship [see **Practicum**].

Journal Club - A group of individuals meeting periodically to study a topic of mutual interest using previously assigned resource materials as a guide to discussion.

Learning Consultant - A person qualified in the development and administration of individualized learning plans which include needs assessment, statement of objectives, selection of learning strategies, monitoring of progress, and evaluation of achievement.

Lecture - A carefully prepared oral presentation of a subject by an expert.

Library Personnel Interchange - Exchange of librarians; an arrangement by which two libraries lend to each other simultaneously one or more staff members for a limited period.

Learn - To acquire knowledge of or skill in by study, instruction or experience; to become informed of or acquainted with; to ascertain; to gain by experience, exposure to example, or the like.

Merit - The excellence of a programme, activity, etc. as assessed by its intrinsic qualities or performance.

Needs Assessment - Systematic process used to identify needs including both gathering and analyzing the information.

Non-individualized Method of Learning - Methods and programmes which provide learning experiences at a pace that lies beyond the learner's control, including the use of tape, radio and television.

Non-traditional Learning Experience - Non-traditional programmes, learning experiences, or methods of instruction are ambiguous terms. They are generally viewed as nearly all programmes, experiences, and methods of instruction which are different from the typical campus and classroom-bound, teacher-led, face-to-face, lecture or seminar-type of instruction.

Objectives - Statements of criteria by which one measures the degree to which the purpose is achieved. The statements are made in terms of results to be achieved rather than methods to be used. Objectives are narrower than goals, are usually short range, and are measurable.

> **Instructional Objectives** - statements that specify the kind(s) of behavior expected to occur in learners.
>
> **Programme Objectives** - Statements that describe the intent, function, and results expected from the total learning programme.

Offering - A continuing education activity or experience presented by or made available by a provider.

Performance Standard - That level of expertise on a given essential performance characteristic that is considered to be minimally acceptable.

Planned Programme of Continuing Education - The concept of a planned programme of continuing education involves the organized presentation of a body of knowledge so that the subject is comprehensively covered in sufficient detail to meet the educational objectives. By using appropriate objectives, very narrow or highly specific, continuing education programmes can be developed so they meet the definition of a planned programme of continuing education. A series of unrelated lectures dealing with different topics and covering only one or two points of each topic would not be a planned programme of continuing education.

Practicum - Actual work by a student in one or several libraries for a definite period as part of a library school course of study or a training course in a library. Field work, field practice, and practice work.

Programme - A coordinated variety of learning activities presented by an organization which may be sequentially planned over a substantial time span and may be directed toward defined objectives. A programme includes many segments which are described as educational offerings, activities, or courses (total continuing education programmes).

Provider [see Sponsor/Provider]

Recognition System - A system established and supported by various occupations to carry out activities such as the following:
1. recognize the individual who has achieved the voluntary requirements of the programme;
2. implement a system of record keeping for continuing education;
3. implement criteria or standards for continuing education;
4. review or assist in the review of continuing education programmes/offerings.

Self-Assessment - An evaluation by an individual of his/her own strengths and weaknesses.

Seminar - A small grouping of people with the primary emphasis on discussion under a leader or resource person or persons. Offers the advantages of guidance and expertise with in-depth discussion. Can be a one-time event or a series over a period of time.

Short Course - A sequential offering, as a rule under a single instructor, meeting on a regular basis for a stipulated number of class sessions over a short period of time (e.g., one to three weeks, etc.). The non-credit course may resemble the credit course in everything but the awarding of credit. It may also be more informal and more flexible in its approach in order to meet the needs of students.

Software - 1. A general term for audiovisual materials that are used in conjunction with hardware (e.g., videotapes, audiotapes, slides, films, etc.). 2. A general term for the programmes and supportive documentation that are used in conjunction with computers and other data processing systems.

Sponsor/Provider - The institution, organization, or agency responsible for the development, implementation, evaluation, and record keeping of a continuing education offering or a total continuing education programme. A Learning Consultant can be considered a sponsor/provider.

Staff Development [see Continuing Education]

Symposium/Forum - Similar to conference/institute. Audience participation.

Telecommunications - The transmission of communication [including educational activities] by electronic means, such as telegraphy, cable, telephony, satellite, radio or television.

Tutorial or **Primer** - A brief remedial exposition of a subject in order to provide a foundation for more in-depth learning; generally delivered by an expert in the lecture mode.

Workshop - Usually meets for a continuous period of time over a period of one or more days. The distinguishing feature of the workshop is that it combines instruction with laboratory or experimental activity for the participants. The emphasis is more likely to be on skill training than on general principles.

Worth - The value of a programme, activity, etc., in relationship to a [stated] purpose.

APPENDIX A

HISTORY

The Programme for Quality in Continuing Education represents the culmination of a series of events, activities, and research efforts.

- **1973.** A nationwide survey of practitioners and of continuing education providers was funded by the National Commission on Libraries and Information Science (NCLIS). (1) This study found serious unevenness in the quality of continuing education and an interest in a voluntary recognition system that include a provision for assurance of quality in programmes.

- **1975.** The Continuing Library Education Network and Exchange (CLENE) was organized during the annual American Library Association Conference in San Francisco at a meeting called by NCLIS who served as a catalytic agent. The first meeting of the Board of Directors was held in September.

- **1976-1977.** A CLENE research project was funded the US Office of Education. A model for a voluntary recognition system was designed. (2) The major elements of a recognition system were identified as 1. criteria and guidelines for quality continuing education single offerings and/or total programmes, 2. a voluntary approval system for providers and recognition of approved providers 3. a voluntary record keeping system for each individual's continuing education and a way of recognizing these efforts, and 4. support services for the individual participants engaged in continuing education such as the development of self-assessment tools, and for the providers of continuing education, such as a handbook to assist in attaining the Criteria for Quality.

- **1977. (October).** The CLENE Board of Directors endorsed the concept of a voluntary recognition system for continuing education and accepted a leadership role in implementing such a system. A Task Force was appointed to develop a plan for the implementation of the recognition system.

- **1978 (January).** The CLENE Task Force presented the concept of a voluntary recognition system at an open meeting at the American Library Association Midwinter Conference. This was followed by many other meetings held nationwide to describe the system and to obtain feedback.

- **1978 (June).** A National Conclave was called by CLENE under the direction of the Task Force. Representatives of the major information, library, media associations and organizations were invited to attend. Conclave participants concluded that there is a need for some type of quality assurance for continuing education activities and programmes.

- **1979-1980.** A CLENE research project was funded by the US Office of Education under Title II-B of the Higher Education Act to develop and demonstrate Criteria for Quality for Continuing Education activities and providers.

The project entailed: 1) forming a seven member Council on Quality Continuing Education: Information-Library-Media Programmes; 2) developing a policy statement on continuing education which includes its definition, objectives and rationale; 3) developing Criteria for Quality for continuing education activities; 4) devising procedures for a voluntary provider approval system for continuing information, library, media education; and 5) conducting pilot surveys of selected providers of continuing education to test the Criteria for Quality as a basis for evaluating and approving programmes.

The Council used the CLENE model of a recognition system designed by the 1976-1977 research project as a basis for its work as well as the criteria and guidelines for providers that have been developed and implemented by other professions such as the American Psychological Association, the American Pharmaceutical Association, the American Institute of Certified Public Accountants, the American Medical Association, and the American Society for Medical Technologists.

1979 (Summer). The Task Force recommended, and the CLENE Board of Directors approved, the adoption of the National Registry System of the American College Testing (ACT) Corporation as means to provide a voluntary record keeping system for individuals who participated in continuing education activities.

1. Elizabeth W. Stone, Ruth J. Patrick and Barbara Conroy, *Continuing Library and Information Science Education.* Final Report to the National Commission on Libraries and Information Science (Washington, D.C.: American Society for Information Science, 1974).
2. Elizabeth W. Stone, Eileen Sheahan and Katherine J. Harig. *Model Continuing Education Recognition System in Library and Information Science.* (New York: K.G. Saur, 1979).

UBCIM
Universal Bibliographic Control and International MARC

UBCIM Publications

ISBD(CF):
International Standard Bibliographic Description for Computer Files

1990. V, 98 pages. Pb. DM 52.00
ISBN 3-598-10983-0

ISBD(CM):
International Standard Bibliographic Description for Cartographic Materials

Revised edition 1987
VII, 55 pages. Pb. DM 48.00
ISBN 3-598-10952-0

ISBD(M):
International Standard Bibliographic Description for Monographic Publications

Revised edition 1987
VIII, 62 pages. Pb. DM 48.00
ISBN 3-598-10953-9

ISBD(NBM):
International Standard Bibliographic Description for Non-Book Materials

Revised edition 1987
VII, 74 pages. Pb. DM 48.00
ISBN 3-598-10954-7

ISBD(S)
International Standard Bibliographic Description for Serials

Revised edition 1988
VIII, 76 pages. Pb. DM 48.00
ISBN 3-598-10955-5

International Guide to MARC Databases and Services
National Magnetic Tape, CD-ROM, and Online Services and Databases

Edited by Wilhelm Marckwardt
3rd rev. enl. editdion 1991
Ca. 250 pages. Bd. Ca. DM 148.00
ISBN 3-598-10987-3

UBCIM is one of the five main programmes of the International Federation of Library Associations and Institutions. The *UBCIM* publications provide the user with deeper information about those bibliographic norms, the development and use of which has become indispensable in the international bibliographic exchange. The publications of *UBCIM* offer a comprehensive and detailed surview of the national bibliographic services.

UBCIM
Universal Bibliographic Control and International MARC

UBCIM Publications

Serial Holdings Statements at the Summary Level: Recommendations

Prepared by Marjorie E. Bloss for the IFLA Section on Serial Publications
1985. I, 56 pages. Pb. DM 48.00
ISBN 3-598-10965-2

Recommended Standards for Cataloguing-in-Publication: the CIP data sheet and the CIP record in the book

Edited by the IFLA
1986. III, 30 pages. Pb. DM 48.00
ISBN 3-598-10964-4

UNIMARC Manual

Edited by Brian P. Holt with the assistance of Sally McCallum and A. Long
1987. V, 482 pages. Pb. DM 78.00
ISBN 3-598-10960-1

Elaine W. Woods
The MINISIS/UNIMARC Project Final Report
A study conducted for the IFLA Programme Management Committee

1988. VI, 18 pages. Pb. DM 38.00
ISBN 3-598-10961-X

Proceedings of the National Bibliographies Seminar
Brighton, 18 August 1987 held under the auspices of the IFLA Division of Bibliographic Control

Edited by Winston D. Roberts
1988. IV, 76 pages. Pb. DM 78.00
ISBN 3-598-10966-0

Standard Practices in the Preparation of Bibliographic Records

Compiled by Dorothy Anderson
Revised edition 1989
VIII, 62 pages. Pb. DM 38.00
ISBN 3-598-10967-9

UNIMARC in Theory and Practice
Papers from the UNIMARC Workshop, Sydney, Australia, August 1988

Edited by Sally McCallum and Winston D. Roberts
1989. VIII, 90 pages. Pb. DM 42.00
ISBN 3-598-10981-4

Manuel UNIMARC
Version française

Traduit par Mireille Chauveinc, Th. Cloarec and S. Jouguelet
1991. IX,419 pages. Bd. DM 68.00
ISBN 3-598-10984-9

K·G·Saur München·London·New York·Paris
K G Saur Verlag Ortlerstrasse 8 Postfach 70 16 20 D-8000 München 70 Tel (089) 76902-0

FSC
www.fsc.org
MIX
Papier | Fördert gute Waldnutzung
FSC® C083411

Kolibri 360 GmbH
Ferdinand-Jühlke-Straße 7
99095 Erfurt, Deutschland
produktsicherheit@kolibri360.de